THE
TENTH
GENERATION

THE JOHNS HOPKINS UNIVERSITY PRESS

BALTIMORE AND LONDON

THE
TENTH
GENERATION

THE ORIGINS OF THE BIBLICAL TRADITION

GEORGE E. MENDENHALL

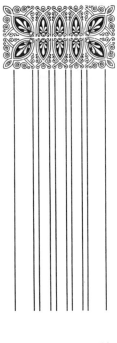

Copyright © 1973 by The Johns Hopkins University Press
All rights reserved. No part of this book may be reproduced or
transmitted in any form or by any means, electronic or mechanical,
including photocopying, recording, xerography, or any information storage and retrieval system, without permission in writing
from the publisher.
Manufactured in the United States of America

The Johns Hopkins University Press, Baltimore, Maryland 21218
The Johns Hopkins University Press Ltd., London

Library of Congress Catalog Card Number 79-189068
ISBN 0-8018-1267-4

Library of Congress Cataloging in Publication data will be found
on the last printed page of this book.

TO
PAUL E. HUFFMAN, *in memoriam*
and
JULIA KOLDEWEY HUFFMAN
who will know why

CONTENTS

PREFACE

For a generation of scholarship there has been very little progress in the development of a consensus concerning the beginnings of biblical history.[1] Radically differing and mutually incompatible views are held with firm conviction. Meanwhile, very rapid progress in the recovery of ancient history and culture through archaeological work and discovery has placed our knowledge of ancient man on a level quite different from that which obtained at the beginnings of modern biblical scholarship. In spite of that fact, relatively little "spillover" into biblical studies has taken place except in details that have little historical significance in themselves. The tendency to treat biblical history as a completely closed system unrelated to the history and culture of the surrounding territories seems endemic to the field.[2] Moreover, there are still many who feel that historical study of the Bible verges on irreverence—that the "sacred" must be exempted from critical investigation.[3] There is, of course, absolutely no reason for this attitude.

1. An excellent recent summary of the situation is found in *LIS*.

2. Compare the long traditional contrast between secular history and "Heils-geschichte."

3. For an extreme example of this attitude—and from a political scientist—cf. "Clarification of the mythical penumbra would, however, only worsen the predicament: Our penetration of the 'truth' experienced by a prophet may destroy it, because for him reality was not only organized a priori but also expressed in terms of mythical configuration. Truth is blinding; stared at too much, it becomes a lie" (J. Meisel, *Counterrevolution: How Revolutions Die* [New York, 1966], p. 38). The complete skepticism concerning the possibility and advisability of obtaining a historical understanding of the prophets expressed in this quotation is derived largely from the fact that Meisel uses such antiquated materials that they have to do only with the history of the discipline of Old Testament studies, especially Max Weber. The old nineteenth-century battle for the freedom of the biblical scholar to follow where evidence leads him was largely won, but for the most part this was at the expense of his scholarship's being regarded as irrelevant to the concerns of religious communities (which of course is to a considerable extent true). It is not the original intent or content of the Sacred Book that is important to religious communities, but its authority to undergird what the community holds at a particular time to be true. For a systematic investigation of this contrast, see J. H. Plumb, *The Death of the Past* (Boston, 1970).

Historical investigation is necessary to understand the meaning of the biblical tradition, and refusal to engage in serious historical study can only be based on a fear that the truth will somehow be harmful to modern man.[4] At a period in the history of Western civilization when the biblical religious tradition is under fiercer attack than at any time since the Roman Empire, a period when the grossest sort of ignorance and misunderstanding of the Bible is something that has to be taken for granted, especially in circles of the educated, it seems imperative that the study of biblical tradition in its own context in the pre-Greek ancient world be pursued with a greater seriousness than ever before.

The validity of the biblical tradition in human experience will not be endangered or destroyed by historical investigation.[5] It is, after all, the only religious tradition of the ancient world that survived in any systematic way.[6] It can be ignored only upon the assumption that the human experience of the past is without value for the present. This may be true in highly technical fields in the sciences, but the humanities do not so easily become obsolete, much as some of its representatives would like so to believe. The present generation, it would seem, has exhibited relatively little talent in the realm of human relationships and operating value systems, and, consequently, can hardly claim any justification for rejecting the past experience of humanity in this all-important field of human history and culture.

The present series of essays represents a preliminary attempt to place biblical history upon a somewhat different ground. It suggests that the formation of the biblical community was actually a process which has all the earmarks of a cultural and religious as well as political revolution. Since revolutions are not well looked upon in religious circles, it is necessary here to describe what is meant. The proper religious term is "reformation," but a reformation is by definition a return to the tradition which is regarded as original, authentic, and true. Though it is virtually certain that in the time of Moses and Joshua there were returns to much earlier tradition, these are, in the present view, merely the incidental, formal details that are inevitable in any radical discontinuity. If the present is intolerable, and the future inacces-

---

4. It is of course not real persons, but corporate bodies that are challenged by such a study. However, the challenge to social power structures is a constant in the biblical history. The heresy trials of the nineteenth (and twentieth) centuries in ecclesiastical circles are dramatic illustrations of the fear of historical truth, or alternative concepts thereof.

5. After all, the first history writing worthy of the label is found in the Hebrew Bible: the Bible represents a radical transition in the foundations of religious faith from ancient pagan myth to historical event. To insist upon the validity *only* of a reduction of history to myth is to assert the truth of ancient paganism, not the Bible.

6. Not only in Judaism, of course, but also in Christianity and Islam.

sible, only the past can furnish models for change that can be communicated to a mass population with relative ease. But, except in certain details of social and technological culture, the biblical community was not a return to the past.[7] What was important about this community was its radically new way of looking at God, nature, and humanity—and this was truly revolutionary. A revolution occurred that is just as relevant today as it was in the time of Moses, and one that is just as necessary.

Implicit in all the essays of this volume is the view that the early biblical community was founded upon a religious covenant, the form and nature of which I briefly described in 1954.[8] Although the study was based upon a formal analysis, many subsequent studies have failed to see that the real importance of the Decalogue-Covenant is not its form but its social and religious function in the formation of the community and the foundations of conviction which determined the entire subsequent history of the biblical religious tradition. Much as it was modified, changed, rejected, and readapted in the centuries following Moses, the covenant tradition nevertheless was what underlay the community's self-understanding until the radical changes that took place during the Hellenistic and Roman Empires made the disintegration of the old tradition in favor of radically new adaptations inevitable.

According to the thesis being presented here, the biblical revolution occurred within the framework of a kind of recurrent dilemma of humanity (chap. IX). Whether or not the thesis corresponds to the historical facts is a problem which may or may not be decided by the accidents of historical and archaeological discovery, but that this approach to biblical history and thought is useful in understanding ancient man and the biblical religious value system can hardly be called into question. To modern man, the biblical language seems hopelessly archaic and quaint in translation; it is embarrassing in its intensely personal and ethical thrust, and in its constant involvement with the experience of God and the "supernatural." Yet it seems certain that the same situation held true in the time of the prophets, many of whom risked their lives to speak effectively for a most unpopular point of view. Like many other mutations in the history of human evolution, the biblical one went "underground," so to speak, surviving only in those individuals who could not simply conform to the insane fashions and conflicts which had to lead to destruction.[9] In fact, it seems valid to conclude that the biblical tradition has

7. A. Glock, "Warfare in Mari and Early Israel" (dissertation, University of Michigan, 1968).

8. G. E. Mendenhall, "Covenant Forms in Israelite Tradition," reprinted in *BAR*, volume 3. For more recent discussion see Delbert R. Hillers, *Covenant: The History of a Biblical Idea* (Baltimore, 1969).

9. Isa. 1:5.

always been most creative and socially functional during periods when the community was without any great political or economic power. Conversely, the most shocking and corrupt periods in the history of religion generally occur when the religious system is identified with a monopoly of force. It is then that the religious and political tribalisms reinforce each other to create the atrocities for which religion is usually blamed.

Religion should be blamed—but it is not the normative biblical tradition that has gained the ascendancy when such atrocities occur. Rather, it is the natural pagan religion of unregenerate man that identifies the distinction between good and evil with an arbitrary political boundary line,[10] and in so doing identifies God with the political monopoly of force. Though virtually all scholars in recent decades have agreed that a radical contrast to ancient paganism was introduced into the history of religion with Moses, the nature of that contrast has still not been made very clear. It now seems that the reason for this failure lies at least partly in the fact that scholars have assumed that early Israel was a typically "primitive" culture until it evolved sufficiently to become a civilized state. Such theories are here assumed to be completely without historical foundation; all of the evidence cited to support them can and should be placed into a radically different context—the inevitable disintegration of the economic system that accompanies revolutions, and the disvaluation of the luxury objects that we use to measure the degree of "civilization" that a particular culture has attained. The constant identification of the wealthy with the unscrupulous and the evil in biblical tradition[11] most probably stems from this formative period of the biblical community, when it was enough for every man to "sit under his own fig tree and under his own grape vine."

Although it is self-evident that these essays are only the beginning and not the last word in attaining a satisfactory historical understanding of the beginnings of the biblical religious tradition, the fact that the early community was far more complex than scholarship has been able to see should emerge as undeniable. Arising out of a worldwide political and economic catastrophe,[12] and at the same time presenting a new religious synthesis based upon the subordination of power to ethic and economic concerns to human relationships, a new social synthesis was created as well. Like all the other world religions, the new synthesis brought about a unity out of diversity, and a

---

10. That this is the real historical content and function of the old mythical cosmic dualism is argued below, chap. VII. For a description of a modern primitive conceptualization that seems strikingly similar, see J. Middleton, *Lugbara Religion* (London, 1960), pp. 233–34.

11. Matt. 19:24; Deut. 32:15.

12. See below, chap, I.

peace where there had been conflict.[13] The basis of this religion was the rejection of control of human beings by force,[14] and the proclamation that only God was in control—through the voluntary subjection of all members of the community to those policies of the sovereign stipulated in the Decalogue-Covenant. This was the constitution of the "Kingdom of God"; that it existed at all is a historical miracle.

These essays are dedicated, then, to the historical understanding of the formative period of the biblical community. Using many different materials and methods, I have attempted to create a new historical context within which the narratives begin to make historical sense. I do not, by any means, reject the canonical history, as did so many nineteenth-century scholars, since the recovery of the past has demonstrated the foundations of the biblical traditions in historical reality far too often.[15] But if the canonical history were satisfactory as a description of the real-life events, there would be no scholarly task. And, as a matter of fact, none but the ultrafundamentalist has been satisfied with the canonical history.

It is for this reason that historical investigation is not only justified but essential to the understanding of the biblical tradition. The forms and formulas survived, but it seems characteristic of religious traditions to become entirely insensitive to "correspondence with reality" outside the ritual observances,[16] or even to object to the idea that rituals ever had such correspondences.[17] The hypothesis that this process of cumulative nonfunctionality of ritual forms was largely responsible for the long-held academic idea that early Israel was a "primitive" society is a most tempting one, since it is precisely this extreme concern for tribal or clan ritual that characterizes such primitive groups.[18] Yet it should now be perfectly clear that the enormous elaboration of cultus, illustrated, for example, in the Book of Leviticus, is a

13. The current and past emphasis upon Yahweh as "warlord" is only one aspect of the fact.

14. There was of course no absolute renunciation of force, but such use was thoroughly subjected to divine control rather than permanently delegated to any person or institution, and was exercised legitimately only against the internal and external enemy of God. See chap. III.

15. G. E. Wright, *Biblical Archeology* (Philadelphia, 1957).

16. For a classical statement, see Isa. 29:13. Conveying to the modern pagan the fact that "religion" does not merely consist of ritual observances is a most difficult task.

17. Typical has been the dogma that the sacred rituals have derived directly from divine inspiration. The suggestion that they originally had some historical and social context seems blasphemous to most religious "conservatives." What is being protected by this attitude is not the original intent or content of the form, but the authority of the socioreligious institution.

18. *PSO*, p. 126.

function not of early Israel but of a very specialized group of high social status existing during the time of the Solomonic temple.[19] In other words, the function of whatever cultic forms there were had made a most radical shift from a concern for functional ethic and morality (the Decalogue) to social status.[20] The correlation between ritual and high status in the community has become a truism in anthropology,[21] and the exclusion of laity and nonmembers of the community from ritual functions or ritual participation is a striking constant in the history of religion, with some exceptions in which the ritual inclusion of foreigners is similarly formalized (see the chapter on Baal Peor).

Though the essays deal with specific historical problems, they all are intended to relate to the common theme of the Imperium of God (the "Kingdom of God" [chap. I]) as a functioning historical reality in the period from Moses until the reversion to ancient paganism which reached its climax in the reign of Solomon. In the chapter on the mask of Yahweh (chap. II), the experiences by which the God was recognized and the phenomena related (through the covenant form) to the establishment of the rule of an extrasocial, that is, "transcendent," factor of ethical content are discussed. The chapter on Yahweh Vindex (chap. III) illustrates the means by which ancient language gave expression to Yahweh's exercise of the Imperium— often, as in any political state, through human or social agency. This essay is the outgrowth of an observation by W. F. Albright many years ago, and I owe him an apology for having postponed so long the publication of the results.

The Baal Peor incident (chap. IV) illustrates the functioning of the covenant as the basis of social morality, and the curse-blessing formula. The two chapters on the ᶜApiru and the Hivvites (chaps. V and VI) are a mere preliminary in a vast problem that has barely been touched, namely, the description of the society itself, its extremely complex relationships to the various cultures in the vicinity, and its internal diversity. The chapter on tribe and state (chap. VII) contributes further discussion on the nature of the early society as a new phenomenon in the evolution of human societies, followed by a counterrevolution—the reversion to an earlier type. Finally,

19. Cf. R. Dussaud, *Les origines cananéennes du sacrifice israélite* (Paris, 1941). Cf. also the various prophetic denials that sacrifice (i.e., the temple system) had anything to do with early Israel, e.g., Jer. 7:21–22.

20. It is quite probable that the redefinition of "religion" found in the so-called "ritual decalogue" of Exod. 34:10–28 must be correlated with the transition from a religious community to a political state.

21. Middleton, *Lugbara Religion*, pp. 71–78. The death penalty for unauthorized ritual acts is frequently mentioned in the priestly sections of the Pentateuch; see now Lev. 12:2; Num. 16, 17:13; and chap. VI, n. 62.

in the discussion of religion and politics (chap. VIII) I attempt to place the issue squarely where it belongs and where it has been placed by the biblical tradition—upon the individual and his willingness to be guided by ethic rather than force.

It is most fitting to paraphrase Paul, "What have I that I have not received?" (I Cor. 4:7). It is hardly possible to give proper acknowledgment to the gifts from a multitude of colleagues and students during the ten- to twenty-year period in which these essays have been growing and taking shape. Ancient man had absolutely no respect for the modern boundary lines between university departments, and I have received counsel and valuable leads from many outside the field of biblical studies. Most important have been suggestions from Professors George G. Cameron, Gernot Windfuhr, Stephen Tonsor, and Delbert Hillers, but it must be emphasized that they are in no way to be held responsible for the conclusions here presented.

Grateful thanks are here extended to the History of Ideas Club of The Johns Hopkins University for the invitation to deliver two of the essays as the A. O. Lovejoy Lectures in March 1967. Other essays were first presented as informal seminar lectures during the winter of 1965–66 at the American School of Oriental Research, Jerusalem, Jordan; at the American University in Beirut, and at Harvard School of Divinity.

By no means least is my debt to the most excellent and intelligent typing of the difficult manuscripts done far beyond the call of duty by our departmental secretaries, Mrs. Darla Ladkau, Miss Veronica Blaschak, and others.

Important segments of the materials presented here would never have seen light were it not for the overseas travel and research made possible by the Rackham School of Graduate Studies of the University of Michigan and the Center for Middle Eastern and North African Studies.

The reader should note that full information for abbreviated and shortened citations appears in the Bibliography.

LIST OF ABBREVIATIONS

AASOR	*Annual of the American Schools of Oriental Research*
ABL	R. F. Harper, *Assyrian and Babylonian Letters*
AH	A. Erman and H. Grapow, *Ägyptisches Handwörterbuch*
AHW	Wolfram von Soden, ed., *Akkadisches Handwörterbuch*
AJ	*Antiquaries Journal*
AJA	*American Journal of Archaeology*
ANEP	J. B. Pritchard, ed., *The Ancient Near East in Pictures*
ANET	J. B. Pritchard, ed., *Ancient Near Eastern Texts relating to the Old Testament*
APN	Herbert B. Huffmon, *Amorite Personal Names in the Mari Texts*
ARAB	D. D. Luckenbill, *Ancient Records of Assyria and Babylonia*
ARI	William F. Albright, *Archaeology and the Religion of Israel*
ARM(T)	*Archives royales de Mari* (traduit)
AT	D. J. Wiseman, *The Alalakh Tablets*
ATD	Das alte Testament deutsch
BA	*Biblical Archaeologist*
BANE	G. E. Wright, ed., *The Bible and the Ancient Near East: Essays in Honor of W. F. Albright*
BAR	*The Biblical Archaeologist Reader*
BASOR	*Bulletin of the American Schools of Oriental Research*
BWAT	*Beiträge zur Wissenschaft vom Alten Testament*
CAD	A. Leo Oppenheim *et al.*, eds., *The Assyrian Dictionary of the Oriental Institute of the University of Chicago*
Capp.	Cappadocian = Hattic, the pre-Hittite language of central Anatolia
CS	Henri Frankfort, *Cylinder Seals*
CTM	*Concordia Theological Monthly*
DLL	E. Laroche, ed., *Dictionnaire de la langue louvite*
EA	J. A. Knudtzon *et al.*, *Die El-Amarna-Tafeln*
EEP	Nelson Glueck, "Explorations in Eastern Palestine," *AASOR*, vols. 14, 18–19, 25–28
EI	Early Iron Age
FSAC	William F. Albright, *From the Stone Age to Christianity*
GN	Geographical name

HE	J. Friedrich, *Hethitisches Elementarbuch*
HH	E. Laroche, *Les hiéroglyphes hittites*
IAAM	Henri Frankfort et al., *The Intellectual Adventure of Ancient Man*
IDB	*The Interpreter's Dictionary of the Bible*
JAAR	*Journal of the American Academy of Religion*
JAOS	*Journal of the American Oriental Society*
JBL	*Journal of Biblical Literature*
JCS	*Journal of Cuneiform Studies*
JEA	*Journal of Egyptian Archaeology*
JNES	*Journal of Near Eastern Studies*
JPOS	*Journal of Palestine Oriental Society*
JSS	*Journal of Semitic Studies*
LB	Late Bronze Age
LIS	M. Weippert, *Die Landnahme der israelitischen Stämme*
LPG	Ph. H. J. Houwink Ten Cate, *The Luwian Population Groups of Lycia and Cilicia Aspera during the Hellenistic Period*
LXX	The Septuagint Bible
MB	Middle Bronze Age
MT	The Masoretic text
MVÄG	*Mitteilungen der vorderasiatisch-ägyptisches Gesellschaft*
NH	E. Laroche, *Les Noms des Hittites*
NPN	Ignace J. Gelb et al., *Nuzi Personal Names*
OK	Theo Bauer, *Die Ostkanaanäer*
PEQ	*Palestine Exploration Quarterly*
PRE	G. Wissowa, ed., *Paulys Real-Encyclopädie der classischen Altertumswissenschaft*
PRU	*Le Palais royal d'Ugarit*
PSO	E. R. Service, *Primitive Social Organization*
PTU	Frauke Gröndahl, *Die Personennamen der Texte aus Ugarit*
RA	*Revue d'assyriologie et d'archéologie orientale*
RHPhR	*Revue d'histoire et de philosophie religieuses*
RSV	Revised Standard Version
SI	Enno Littmann, *Safaitic Inscriptions*
TET	S. A. B. Mercer, *The Tell El-Amarna Tablets*
UT	Cyrus H. Gordon, *Ugaritic Textbook*
VESO	William F. Albright, *Vocalization of the Egyptian Syllabic Orthography*
VT	*Vetus Testamentum*
ZAW	*Zeitschrift für die alttestamentliche Wissenschaft*
ZKP	L. Zgusta, *Kleinasiatische Personennamen*

I

EARLY ISRAEL AS

THE KINGDOM OF YAHWEH:

THESIS AND METHODS

Quite consciously, the guiding hypothesis has been chosen that early Israel until the establishment of the Monarchy actually constituted the sovereignty of Yahweh as the functional system that enabled it to exist at all as a distinct social group.[1] It is true that such a social system seems quite remote to modern man, but historical analogies in diverse times and places do exist, most notably, the society of Upper Egypt contemporary with the biblical period of the Judges, where the state also seems temporarily to have "withered away."[2] It is also rather tempting to cite the example of the papacy at Rome after the collapse of the Western Roman Empire, although the specific historical circumstances were of course radically different: they always are.

It follows inevitably that the point of view deliberately adopted in these essays represents a fairly radical break with some past traditions about biblical history, including some concepts of the past incorporated into the Bible itself. Whether or not the break is justified the reader must decide for himself on the basis of the evidence and the arguments presented. It need only be pointed out that religious traditions have rarely, if ever, been really concerned with knowledge of what actually happened in history.[3] The historical concern has been rather with the use of history as a ground for

---

1. Compare, as early as 1953, John Bright, *The Kingdom of God* (Nashville, 1953), but with primary attention to the concept and its meaning for the modern church, rather than, as here, the ideological base for a historically functioning social organization and the beginnings of a new cultural synthesis.

2. A. Gardiner, *Egypt of the Pharaohs* (Oxford, 1961), pp. 304ff. Thebes under Ḥriḥor, the priest, is described as a theocracy. The old temple states of Mesopotamia are certainly relevant here too.

3. Plumb, *Death of the Past*, Introduction and chap. 1.

some aspect of the well-being or validation of the religious community, its structure, liturgy, or ethos.[4] The urge for "relevance" must nearly always result in a concept of history characterized by a systematic anachronism, and one result has been a radical division between the religious and the academic concepts of history.[5] However, the City of God is not always wrong, and academia, which is also dominated by temporary intellectual fashions, is not always right.

If these essays serve only to make plausible the thesis that the historical study of the Bible is still in its infancy, they will have served at least one useful purpose. Hopefully, this thesis does not unduly denigrate those scholars of past generations whose work must still be presupposed in many areas of scholarly endeavor. There is a vast difference, however, between the history of a human religious community and the history of a book, and the primary concern of the present essays is the former not the latter. Here lies perhaps the most important contrast between the present and future task of scholarship as opposed to the "literary criticism" of the past which many scholars still feel is the only reliable means of arriving at knowledge.[6]

It is not necessary, however, to engage in the old theological battle about the priority of the community versus the priority of the book. For it is now quite clear to me that both community and book came into existence and became a historical reality as a response to something that happened in history and was therefore prior to both. It is that experience to which human beings responded that constitutes the subject matter of history, and no one singlehandedly creates his own experience except the totally mad man and

4. More typical of ancient paganism has been myth as such a "ground of Being." Cf. Bronislaw Malinowski, "Myth in Primitive Psychology" in *Magic, Science and Religion* (1948; reprinted, Garden City, N.Y., 1954), pp. 100–101, 107. Also, see C. Jung and C. Kerényi, *Essays on a Science of Mythology* (New York, 1949), p. 8: "But what one *does* by means of mythology when one allows the 'telling of myths' to function in the involuntary service of a human community, that is far from being the idle invention of explanations. It is something else. The German language has the right word for it: *begrunden*." When the authors go on to explain (!) myth on the ground of alleged "archetypes," it is only logical to classify that (rather popular) tendency together with equally absurd attempts to explain every difficult episode in the Bible as being due to ancient invasions from outer space—an explanation which is remarkably fashionable among engineers and physical scientists. This merely illustrates the profound need for the security of some "scientific" ground which does not demand an ethical choice—merely the choosing to accept the myth—even among sophisticated intellectuals. At the same time, it expands the authority and jurisdiction of the particular discipline by placing into its competence the explanation of important aspects of human culture, since it a priori excludes all other explanations—*especially the historical or empirical.*

5. Plumb, *Death of the Past.*

6. See *LIS*, and my review in *Biblica* 50: 432–36 (1969).

the over-precious artist[7] who is incapable of understanding any criterion of judgment other than "originality" (which usually consists merely of some formal contrast to that which had already been done). It was the recognition of this fact that prepares the way for the beginning of comprehension of that vivid reality of God that places modern man in such contrast to ancient man. On the other hand, after nearly two thousand years of theology, God has become so intellectually rarified a concept, so gnosticized, that radical theologians have felt the need to proclaim the death of God. Actually, it would be much more to the point to proclaim the death of a theology that has lost contact with the sort of historical, empirical reality upon which the biblical faith was based in the first place.[8] The various proposals that have been suggested for solving the religious crisis of contemporary society constitute either the utter rejection of the biblical tradition by withdrawal from reality in favor of some sort of solipsistic mysticism, or the rejection of the biblical tradition in favor of the Late Bronze Age identification of religion with political power structures.[9] The misunderstanding of the Bible that derives from an inability to understand it in a historical context is a luxury that the religious tradition can no longer afford. It is for this reason that an attempt has been made in these essays to show that the understanding of biblical language in a historical context—its own—may at least serve to correct the myriad of distortions, misconceptions, and outright untruths that are still being perpetuated—particularly by many intellectuals who know nothing of ancient Near Eastern cultures, languages, and history. On the other hand, much of theological writing for the past century has been dominated by the unquestioned assumption that, as a mere record of primitive beginnings, the Old Testament is either irrelevant to Christianity or of purely negative importance to show what Christianity is not.[10] Though it is

7. Review of Susan Sontag, *Styles of Radical Will*, by Julian Mitchell, *Manchester Guardian Weekly*, December 13, 1969: ". . . If she ever showed some awareness of what was going on outside her own skull."

8. The "political theology" of Harvey Cox (*The Secular City* [New York, 1965], p. 254) is an excellent summary of what I have tried to describe as the essence of Baal-worship in the ancient world: "It [i.e., politics] brings unity and meaning to human life and thought." This is a valiant act of faith in the art of manipulating the monopoly of force.

9. Therefore, states and ideologies which regard the political power structure as the supreme concern must be either contemptuous of or bitterly hostile toward competing religious systems, particularly the Bible.

10. Well described for the German scene by H.-J. Kraus, *Geschichte der historisch-kritischen Erforschung des alten Testaments* (Neukirchen, 1956), pp. 382–94. The issue at the present time is not the relevance of the Old Testament. It is the relevance of the entire Bible, particularly in religious organizations, that is the problem at present, at a time when religious leaders are engaging in frantic efforts to conform to the changing fads and fashions of the young in order to maintain their leadership.

not my purpose to argue the case for a reevaluation of the Old Testament tradition, I should perhaps point out that it would be advisable to understand it before it is evaluated. If the conclusions in the essays to follow are even in part faithful to the ancient realities of life and thought, then the conclusions drawn from prior misunderstandings can no longer rationally be held.

It is generally accepted that the methods used in any field grow out of the nature of the evidence itself. The long traditional method of literary analysis is hardly a historical method at all; at best, all it can yield is a history of the narrative, as I have argued elsewhere.[11] Dating a biblical passage to a particular time has relatively little to do with the process of evaluating historical evidence. Similarly, the isolation of a particular literary form has no necessary relationship to evaluation and utilization of historical evidence.[12] Furthermore, form criticism has the great disadvantage of being far too hypothetical and theoretical to be satisfactory as a historical tool. One must always ask whether a hypothetical literary "form" which the literary critic extracts from the written materials was really a distinct functional "form" to ancient man, or whether it is accepted as one merely because it happens to correspond to some structured concept or pattern of thought to which modern scholars are at least temporarily sensitized.

Where evidence is minimal, as in most of biblical history, assumptions and hypotheses consciously or unconsciously held are likely to become very powerful determinants of conclusions.[13] Any study or discussion of methodology must begin with this problem. Again, a hypothesis consciously adopted is less misleading than one that is uncritically assumed.

Method begins, then, with the critical examination of widely held assumptions; it is this that has induced me to reject many of these assumptions as untenable. Frequently, they are modern word labels that lent themselves to the biblical materials in the nineteenth century when the "Bedouin mirage" seemed to explain everything important about the puzzling aspects of biblical history. It was assumed then that the early Israelite tribes were nomads in the process of sedentarization,[14] but now it seems most improbable that the Bedouin type of nomadism even existed in the Late Bronze Age.[15] The gap between the Bedouin culture and the ancient realities of life is

11. *BANE*, pp. 32–53.

12. The history and social function of the form is also necessary.

13. Theological disciplines are notorious for continuing to adhere to outmoded generalizations or hypotheses long after they have been abandoned in ancillary disciplines such as philosophy, psychology, sociology, and history.

14. *LIS*, pp. 102–23, and most of Old Testament scholarship at present.

15. *ARI*, pp. 96–97. He still has not fully seen the implications of his own demonstration. See J. Tracy Luke, "Pastoralism and Politics in the Mari Period" (Ph.D. diss., University of Michigan, 1965).

now usually glossed over by the term "seminomad," which is not subject to any cultural analysis or description. It merely serves to perpetuate those aspects of the Bedouin mirage that have become grafted onto modern ideas about biblical history.[16]

Another assumption that seems never to have been seriously questioned is the anachronistic reading back into remote periods of the ethnocentricity of Ezra and Nehemiah and the subsequent postbiblical Judaism.[17] This is closely related to the preceding assumption, since nomadism and "tribal" society have often been equated with each other. Yet the question of what a tribe actually is rarely has been asked other than to answer it on the analog of Bedouin tribes. Finally, when the theologians take over, the Old Testament religion is dismissed as "tribalism" which is incompatible with the modern world; in view of the fact that early Israel consisted of twelve tribes, not one, it is clear that what the theologian dismisses is not the Old Testament, but his own erroneous ideas about it. Quite the opposite conclusion is much more appropriate to the materials incorporated in the biblical writings, that is, that the early faith of Israel consisted historically of a transcending of tribalism. On the other hand, since the kinship system of at least some of the ancient tribes of Israel consisted of the well-known five-generation pattern,[18] the foundation for large-group solidarity could not conceivably have been kinship.[19] Further, it is quite clear from ancient materials that kinship terms were characteristically appealed to as a means for expressing an existing social solidarity that was actually produced by something else.[20] Freed, then, from the uncritical acceptance of late ideas about early Israel, one can ask historically relevant questions: What did bring about the solidarity of the twelve-tribe society? What is the social function of a "tribe" and how does it operate? Why did so many of them cease to function historically at a fairly early point in biblical history? We do not have adequate evidence to answer all of these questions with confidence, and any

16. Roland de Vaux, "Nomadism and its Survival," *Ancient Israel: Its Life and Institutions* (New York, 1961), pp. 1–15.

17. In spite of Albright's own observations of the very complex make-up even of populations in the patriarchal period (*ARI*, p. 99). There is no pre-Mosaic social unity of which Israel is the socioreligious continuation. On the other hand, there is virtually no pre-Yahwistic culture or language that does not have some formal continuity in some aspect of early biblical traditions. See also *PSO*, pp. 176–77: Service's concept of the origins of new cultures contrasts sharply with the historical evidence; the function of time as the integrative factor is here grossly *over*estimated.

18. Presupposed in the Decalogue formula: "to the third and fourth generation . . . ." (Exod. 20:5).

19. *PSO*, p. 111. It is not even the basis of a "primitive tribe."

20. *Ibid.*, p. 131. Ancient political relationships were regularly expressed as kinship terms; see further chap. VII.

answer will be based upon analogies.[21] A major principle for historical study must be, so far as it is possible, to derive analogies from what we know to be true of the ancient Near East, not from modern or postbiblical societies.[22] This procedure is historically necessary, but because it collides with some traditional and modern ideas, it must be discussed further.

Borrowing of forms is an inevitable fact of history, and constitutes a major source for our knowledge of it, but this is not to deny the uniqueness of the biblical society. Quite the contrary, every society is unique, but meaningful forms of human behavior in society cannot be unique.[23] When a new society is in the process of formation, then above all are its forms of communication, organization, and behavior a priori likely to be those that are already well known, even if they are often archaic and nonfunctional historically at the time they are borrowed from the past.[24] The study of forms still does not constitute either history or religion. For such purposes it is necessary to delve beneath the forms of language, organization, and behavior to their functions: the sum of their relationships to other aspects of society and, above all, to the hierarchy of values upon which the social solidarity ultimately depends. It should be clear from the outset that such relationships between form and function cannot be constant in any society except at the expense of complete stagnation. This problem is a constant in law as well as in religion.[25] It is also, however, the answer to the dilemma stated above: the demonstration that forms have been borrowed from outside sources by no means calls in question the uniqueness of the borrowing society, simply because no form can have an identical function in two different societies—either in time or in space. It is for this reason that

21. All history is comparative; it consists in the assumption that there is sufficient comparison between modern language and ancient reality to convey something of the truth to modern understanding. Further, all science is similarly based—for the first activity of scientific endeavor has characteristically been to classify the phenomena on the basis of comparisons—and contrasts.

22. The many jokes about the nineteenth-century German university professor whose blithe assumptions that ancient prophets must think, act, and have the same value system as himself are too true, even today, and not only of university professors.

23. The unique word in the Hebrew Bible is nearly always a problem. To be meaningful, it must have a multiplicity of contexts. When creativity is identified merely with novelty (as it often is in the arts today), it can hardly have any meaning beyond its modern social context, usually functioning merely to attract attention, publicity, and therefore, hopefully, money. It is no wonder that art systematically renounces meaning, and therefore social context, as an aim.

24. Cf. the root NQM (chap. III), or the root agapaō in the New Testament.

25. B. N. Cardozo, Selected Writings, edited by M. Hall (New York, 1947), p. 257: "We may think the law is the same if we refuse to change the formulas. The identity is verbal only. . . ." The essay was first published in 1928.

jurists rightly challenge the idea that law codes are "borrowed"; [26] language may be borrowed, but the functioning law system cannot be. As in every other case of cultural borrowing, there will inevitably be a functional shift. That is, the form will have a range of relationships and meanings that differ from those of its original habitat. The question of the meaning of a form is probably one which is peculiarly characteristic of the philosophical orientation of Western-educated man. As in the famous case of the Navaho borrowing of the Hopi healing ritual, the meaning of the chant's words was so unimportant that the Navaho didn't bother to learn it. The important thing was its efficacious operation—its function. To dismiss this illustration as "primitive" contributes nothing to our understanding and confuses the issue. Usually, the modern pill-taker is not interested in the meaning (or mode of operation, or chemical constituents or possible side effects) of the pill either; he is interested merely in the temporary enhancement of his well-being. It is only the intellectual who is primarily interested in the elaboration of theory or meaning, for this is his role in society.

Unfortunately, addiction to theory is also habit-forming and frequently leads to the subordination of all other concerns to neat intellectual or logical symmetry.[27] Then theology becomes law, and procedures are taken against those who deviate from official theoretical pronouncements.

If we approach this sort of situation, which has characterized all ideologically based societies at some time or another, from an operational, functional point of view, then we have a fairly accessible analogy for understanding the social function and operation of myth in the ancient societies. The problem of demythologizing then appears in a radically different light, for there can be little doubt that the primary social function of the most important myths in antiquity was simply to indicate and promulgate the ultimate metaphysical legitimacy of existing social and power structures. On the other hand, the modern intellectual distrust of ideology misses the point, for it is not ideology that creates social conflict, and such conflict is likely to exist even in the absence of any ideological system. The human obsession for being in control finds expression in many forms other than ideology, and the latter functions primarily as a means, more or less rationally

26. I believe Koschaker somewhere argues this point, but I have not recently found the relevant passage in his writings.

27. Compare the neat, logical theological systems of the past with the concern of jurists for the "*elegantia juris*, or logical cohesion of part with part," mentioned by Oliver Wendell Holmes in *The Common Law* (Boston, 1881), p. 36. "'Talk of stubborn facts,' says Dr. Crothers, 'they are but babes beside a stubborn theory'" (Roscoe Pound, *The Spirit of the Common Law* [Boston, 1921], p. 79).

examined, by which to legitimize actual conduct.[28] In contrast, the non-ideological man often refuses even to examine the assumptions underlying his behavior patterns, or turns the job over to his psychiatrist. The assumption is far too often the unquestioned conviction that he has a right to anything he wants, no matter at whose expense.

In order to illustrate by means of a concrete historical problem, we turn to the form of the Sinai Covenant, which has been much discussed. The thesis presented here is that in the formative period of religious communities, if anything is to be communicated at all (either by action or by words), forms are necessarily borrowed from the past. If this thesis is correct, then it follows that the borrowing takes place by some principle of selectivity that accepts some and rejects others. Furthermore, it seems a priori extremely improbable that at the formation of the community a comprehensive code of morals, doctrines, and liturgies could have been drawn up. One reason is that the community could hardly have been faced with the concrete situations that demanded decisions in all their variety during the first generation, and legislating for the future is a concept that historically does not commend itself.[29] Such theorizing about future possibilities is a function of Greek philosophy,[30] which is the converse of the ancient process of projecting fairly recent norms into the remote past in order to give them authority.

Selectivity then proceeds upon the basis of something other than a prefabricated law code, and, indeed, if such did exist, it would itself be a reflection of this process of selectivity. We probably do have an example of this at a later period in Exodus 21–23. The contrast of its function—what it prescribes and proscribes—to those of the earlier law codes to which it is clearly related,[31] formally illustrates the selectivity; it may be possible to describe the contrast conceptually in modern language. Historically and functionally, the selective principle exists only in the actual act of choosing in

28. Modern mass media, of course, extend to a frightening extent the possibility of obtaining mass acceptance of the behavior not only of manufacturers (television commercials) but also of political parties and states. The enormous danger of this had already been pointed out in 1949 in *Law and Contemporary Problems* (Durham, N.C.: Duke University, 1949), p. 1.

29. It is now law that legislates for future situations; this is one difference between law and covenant—for the stipulations of the covenant were such that social and cultural changes made very little difference in the relevance of the covenant obligations. Law is based upon the accumulated experience of the community, which includes experience with legal doctrine and procedures of the past.

30. Note especially the enormous elaboration of future unreal conditional clauses in Plato's dialogues. Such clauses hardly exist in the ancient Semitic languages except under very unusual circumstances, such as when Jacob predicts his *own* reaction to a *hypothetical* future condition—the death of Benjamin in Gen. 42:38.

31. E.g., the Code of Hammurabi.

a concrete situation.[32] The ground for that selectivity is the heart and substance of religion, of obedience to God, but the formal statements are the
product, not the cause of obedience. It is for this reason, then, that the
"religiousness" of the Covenant Code is hidden;[33] it was not until much
later that religion became identified with form. In early periods religion is
identified with function: what a form does in society and, especially, the
effect of a form of behavior upon other human beings. At the grass roots
level of religion, behavior is often not overtly religious in a formal way,
except after a long process of cultural accumulation of traditions as well as
forms. Then, characteristically, a "liturgical movement" attempts to sanctify
every important facet of life with formal prescriptions.[34] This cultic-
liturgical definition of religious life is the end point of religion, not its
beginning. At the same time, it is true that the formal-liturgical definition of
religion depends for its authority upon the appeal to remotely past precedent.[35] Such formal reductionism can be historically quite useful, and for
those whose sensitivity in religion is predominantly formal, the evidence for
the antiquity of particular forms is evidence of the antiquity of the written
documents that describe them.[36]

Again, the contrast is function. *Si duo idem faciunt, non est idem*, is the
first law in the history of religion. It is the first law in the history of law as
well.[37] Participating in the Christian sacrament of the Lord's supper cannot
conceivably have the same social function now that it did in the early Roman
Empire when the very term *sacramentum* meant primarily the soldiers' oath
of loyalty to the emperor. To affirm that the modern sacrament is identical
with the New Testament one is the height of absurdity, for the relationship
to social reality cannot be the same; indeed, it may be quite the opposite —

32. "The law in books and the law in action are two different things." The source
for this statement has been lost to me.

33. A. S. Diamond, *Primitive Law* (London, 1950), pp. 102–25. The whole treatment is archaic (dating originally from 1935), but is the generally prevailing opinion.
Cf. *Interpreter's Bible* (New York, 1952), I: 844 (especially Coert Rylaarsdam on the
"earlier" part of the Code). H. Cazelles is quite correct: "La couleur religieuse de
l'oeuvre israelite brille toutefois d'un éclat particulier" (*Études sur le code de l'alliance* [Paris,
1946], p. 184).

34. A syriac liturgy provided a ritual that was intended to guarantee that a cow
would not switch its tail into the milker's face. It strikes a responsive chord in the
heart of anyone who has ever milked a cow.

35. Best illustrated in Egypt (see J. Wilson, *The Burden of Egypt* [Chicago, 1951],
pp. 294–96, 305ff).

36. This is of course the foundation of Y. Kaufmann's well-known thesis (*The
Religion of Israel* [Chicago, 1960], p. 175). However, with literary documents as with
archaeological strata, the document is dated by the latest content, not the earliest.

37. Cf. the quotation from Cardozo above, n. 25.

the affirmation of conformity to a dominant social system, or at least to the social system of a traditional group with which the individual identifies himself.[38]

The importance of the element of "tradition" in our present discussion ought to be stressed here. One scholar has recently defined religion as the "cumulative tradition" plus "faith."[39] But no religion which has a point of origin in time and space *starts* with a cumulative tradition; on the contrary, it starts with a profound conviction, which determines behavior, with regard to good and evil. It represents a break with tradition, a discontinuity of existing social and religious systems,[40] even though there can be no absolute cultural discontinuity of forms. In the past, the discontinuity from the Late Bronze Age to the Iron Age has been explained on the basis of a hypothetical change or displacement of population: the Israelites displaced the Canaanites in part, the Phoenicians displaced the Canaanites elsewhere; the Arameans displaced still more, and so on down the line. All of these ideas are now untenable. If the Phoenicians are merely the continuation of Canaanite culture, with considerable changes of course, the Israelites also represent such a continuation with a change of a more radical sort (particularly in the religious and social system). As revealed by excavations, certainly it is true that there are only minimal differences between the two in material culture,[41] and those differences are most readily explained as functions of the differences in the social, economic, and religious structure of the ancient Israelites.

On the other hand, it is also quite clear from Late Bronze Age documents that no culture of the ancient world that yields written documents is ethnically homogeneous.[42] From the beginnings of writing, almost by definition, the city has been a cosmopolitan community. From the beginnings of writing, the identification of all sorts of social terms as "pure" ethnic groups is the sort of modern (or late antique) myth that needs to be

38. The *sacramentum* of the Christians before Constantine must constantly have been interpreted as a conspiracy: a loyalty to someone other than the emperor—which it of course was but in a different dimension.

39. W. C. Smith, *The Meaning and End of Religion* (New York, 1963), p. 156, and the following discussion to p. 192.

40. Perhaps best illustrated in "You shall have no other gods . . ." (Exod. 20:3), and "Choose you this day whom you will serve . . ." (Josh. 24:15). A similar discontinuity is in Matt. 5: "But I say unto you . . ." In all three cases the break is made with the *immediate* past or present. In the case of Jesus, it is in favor of a return to the more remote past: the Kingdom of God, which is about to break into reality for the second and final time.

41. It is difficult, if not impossible, on archaeological grounds alone to determine whether a given site is Israelite or non-Israelite.

42. For Ugarit now, see *PTU*. It is typical, on a large scale, of all our sources from other sites.

given up as soon as possible. Not only does the Bible itself insist upon a common origin of all mankind[43]—the significance of which has too frequently been lost in an age obsessed with scientific precision of statements—it also states flatly the Canaanite, Aramean, and other origins of various segments of the twelve tribes.[44] Finally, the incorporation of the old Canaanite city states under King David[45] has left behind virtually no hint of the population contrast that was thus unified under a common state,[46] particularly for the city of Jerusalem itself.

Ethnic solidarity is essentially a matter of allegiance to a tradition in a continuity;[47] what produces and maintains that allegiance is perhaps the most important mystery of history. It is even very difficult to explain the withering away of it, as in the deaths of ancient Babylon, Assyria, Phoenicia, and Egypt, to name only a few of the myriad extinct cultures of the ancient world. However, the observation of the allegiance to a tradition at a given time must be kept quite distinct from the historical problem of origins, even though the allegiance tends always to be based in a religion that appeals to the past as the ground and precedent for that which *is* at some particular later period. This is why the historical study of a religion is always a sensitive issue, especially for conservatives. Yet such study cannot be subversive of what is of primary importance to the religion: the formative period is what survived and was probably responsible for the creation of the particular value system that enabled the community to exist, no matter how distorted and readapted it may have become in the course of subsequent history.[48]

The view of the past is perhaps the most important contrast between religious communities in the formative period and those content with the mere reception, preservation, and transmission of a tradition. A formative period is by definition one which is concerned to *break* with the contem-

---

43. Gen. 1–2; Luke 3.

44. 1 Chron. 2:3, 7:14; Num. 11:4, to say nothing of the many narratives of individual marriages outside the religious community, especially the early kings.

45. Beginning with Jerusalem itself, 2 Sam. 5, the Philistine cities, 2 Sam. 8, then the empire, 2 Sam. 8. Megiddo, Taanach, and Gezer are all under Solomon's administrative system.

46. This is probably the historical context for the Deuteronomic law providing that interest-free loans are available only to Israelites (Deut. 23:19–20). Other "laws" in Deuteronomy presuppose the same sort of political symbiosis in private life and transactions, and the entry of foreigners into the religious community (Deut. 15:1–3; 23:3, 7–8).

47. *PSO*, pp. 176–77, n. 17.

48. Such distortion is certainly referred to in Jer. 8:8: "the lying pen of the scribes," but the passage is obscure. There is little reason to doubt that much of the injustice and perversion of the law that most of the prophets bitterly condemn was done on perfectly "legal" grounds.

porary and recent past,[49] partly because it is intolerable or unsatisfactory, but more importantly because there comes about a vision and conviction that something much more excellent is not only possible but necessary. Discontent movements are a constant, as the history of revolt, war, and rebellion indicates. But rare indeed are those movements in history that result in such creative breaks with the past that they survive for centuries and expand over large population areas to create some sort of social unity or unified tradition that did not exist before. The first such movement to survive was the biblical one, and this religious revolution urgently needs more adequate *historical* study. It is, after all, a primitive community that has a minimum of remembered history; it takes maturity and honesty to accept and evaluate historical facts as they happened, rather than to gloss them over with such a patina of sentiment and anachronistic rationalization that the events are no longer recognizable. The context of historical reality is lost, and the result is a narrative that so emphasizes the supernatural events that modern scholars have for generations concluded that there is little or no historical foundation for the narratives at all. This is particularly true of those who still adhere to the nineteenth-century ideas of progress, which must posit a very primitive sort of society for the early period of Moses and the Judges.[50]

It is true that the period of the Judges provides us with very little knowledge of some of the most useful criteria for distinguishing between a primitive society and a civilized culture—namely, the division of labor, specialization in occupations, and high degree of economic organization. It must be remembered, however, that the period in question was a "Dark Age" all over the civilized world after the catastrophes that brought an end to the great empires of the Late Bronze Age. Furthermore, particularly in ancient Israel, there was a systematic, ethically and religiously based, conscious rejection of many cultural traits of the Late Bronze Age urban and imperial cultures.[51] There is a vast difference between a society that has never known "civilization" and one that—partly through deliberate choice and partly through circumstances beyond its control—has reverted to a culture of farmers and shepherds. The same thing occurred in the Greek cultures after the collapse of the Minoan-Mycenean Empires.[52]

It is within the context of such vast historical changes that the genesis

---

49. Josh. 24:15, and in this context compare the old narrative of Ruth 1:16.

50. M. Smith, "The Present State of Old Testament Studies," *JBL* 88 (1969): 19–35. Like all Old Testament scholarship and most anthropology (cf. *PSO*, p. 172), the old nineteenth-century contrast between "primitive" society and the "civilized" state dominates the discussion.

51. Kingship, art, professional military, temples, to name a few.

52. T. B. L. Webster, *From Mycenae to Homer* (London, 1958), p. 136: "... and the Greek world became isolated, poor, and fragmentary."

and development of the early biblical faith must now be viewed. The picture that emerges is poles removed from the rather primitive academic tradition of bygone generations which had at their disposal an extremely limited range of social and cultural types and historical evidence within which to view the early history of Israel. It now seems possible to understand the extraordinarily sophisticated religious thought that always constituted the real talent of the biblical community,[53] whereas the areas usually held in highest esteem by ancient and modern paganism such as the arts and architecture, social and political organization, economic structures, and technical bodies of science or pseudoscience, were either virtually nonexistent in ancient Israel or thoroughly derivative.[54]

For the same reason, it is quite clear to the present writer that any sharp separation or antithesis between faith and history,[55] or even any attempt to describe either in isolation from the other, can only result in utter failure or a grotesque distortion of both. In view of the exceedingly involved elaboration of discussion *about* history,[56] it is necessary here only to state that history and historiography are only accidentally related to each other. By history, I mean merely the sum of human experience. History is related inseparably to faith, not only in the religious convictions that determine the selection of data and the conclusions of the one who writes history—this is a relatively insignificant aspect of history, even though it often creates great problems for later historiographers. The most important relation between faith and historical reality is in *acting* itself: the conscious or unconscious value judgment that, on the one hand, determines the actor's behavior, is of the very essence of biblical faith (obedience) and constitutes the real subject matter of ethic and Torah. On the other hand, faith is also absolutely bound up with the *evaluation* of events over which the individual or even the community often has no control. The event and its impact upon the emotions and total well-being of the individual and community are a historical unit, even though the sources often describe the latter much more fully than what the modern historian would consider as the event.[57] This

53. Deut. 32 alone would be sufficient to support this assertion. It is most difficult to avoid placing it into the context of the fall of Shiloh, about 1050 B.C.

54. However, it is clear that by the end of the Late Bronze Age, such aspects of material and social culture were thoroughly derivative almost everywhere else. It was, after all, an international period.

55. R. Pfeiffer, "Facts and Faith in Biblical History," *JBL* 70 (1951): 1–14.

56. Especially in theological literature, see James Smart, *The Divided Mind of Modern Theology* (Philadelphia, 1967), *passim*, and the Byzantine tortuosities of the New Frontiers in Theology series edited by J. M. Robinson and J. B. Cobb, Jr. (vol. 1; New York, 1963).

57. This is particularly true of the literature of Exile from Ezekiel to Deutero-Isaiah, and, according to the present writer, Job.

constitutes one of the most important historiographical problems for the biblical historian, and necessitates a much more subtle method than has generally been used.

The method is readily analogous to that of intelligence work in war-time: that is, the synthesizing of a total picture from bits and shreds of evidence, which becomes possible at all only because of a much more adequate knowledge of the whole historical context of the period under consideration. Further, reconstructing the picture will be possible at all only by systematic rejection of the idea that the formative period of Israel represents a totally new culture unrelated to anything in the past. The old idea of one ethnic group's moving in to displace or destroy a predecessor which then promptly disappears is based upon most unsophisticated notions that cannot be too thoroughly repudiated. Such an interpretation is related to the original function of Flood stories in many cultures:[58] a way to explain a gap resulting from failure of historical memory or due to a period of accelerated social change. With more adequate hypotheses, methods, data, and knowledge of ancient patterns of thought, one can describe the synthesis of faith, action, and experience that constitutes what we now call religion.[59] There is no known symbol in the entire Bible that brings these three together in any satisfactory way other than the covenant form as it has been isolated on the basis of contemporary political forms.[60] In spite of some historical difficulties (or rather difficulties in the sources), the majority of which are shared by the extrabiblical sources, the thesis that the Sinai covenant was actually such a suzerainty treaty has proven to be a useful working hypothesis that can well take precedence over other ideas until something better arrives. Many of the objections to it constitute defenses of existing presuppositions which are themselves merely accepted and not examined.[61]

Since the fundamental function of a covenant was to establish a community of interest between a suzerain and his vassals,[62] it was the covenant

---

58. Both Greek and Latin historians often begin history or historical continuity with the "Deucalionic Flood." Note also Webster's remarks on the poet's use of Eastern concepts of the gods' hostility to man (*From Mycenae to Homer*, pp. 86, 180).

59. I would agree in principle with Smith in *Meaning and End of Religion* that "religion" as now used in academic and popular circles is an artificial concept constructed in the nineteenth century.

60. G. E. Mendenhall, *Law and Covenant in Israel and the Ancient Near East* (Pittsburgh, 1955); reprinted in *BAR*, vol. 3.

61. Cf. Dennis McCarthy, *Treaty and Covenant* (Rome, 1963), and my review in *JAAR* 45 (1967): 303–4. Also, E. F. Campbell, "Sovereign God," *McCormick Quarterly* 20 (1967): 7.

62. The use of kinship terms is most vivid illustration: a suzerainty treaty normally creates a "father-son" relationship, while a parity treaty yields a brotherhood. The family is the community of interest par excellence. In this context, the passage in John

alone that distinguished between a group which must be dealt with by force (war), and a group which could be dealt with according to what we consider as normal, orderly, peaceful procedures. This contrast was something to which society was already very sensitive in the Mari period. The absence of a covenant meant the nocturnal execution of members of a tribe who were being held as hostages.[63] It is a matter of semantic course, then, that the covenant was itself called a *salimum*; a "peace." Over a thousand years later, this contrast was taken so for granted in ancient Israel that it was described as the difference between brute animals and intelligent, understanding man: "Be not as the horse or the mule . . . which must be restrained with bit and bridle lest they attack. . . ."[64] Similarly, the Assyrians described the utter barbarians (from their point of view, of course) as those who know not oath and covenant.[65] It follows that violation of a covenant is tantamount to its absence, particularly in the biblical tradition.[66] This fact—which determines the change in status from a community dealt with by the sovereign's concern for his own property ("holiness") to one dealt with by force and hostility—accounts for the great emphasis upon the wrath and "vengeance" of God in the Old Testament. The unbelievably vicious distortion of this in the history of religion and theology alike can hardly be overstated. Actually, nothing better illustrates the radical breakthrough that the early biblical tradition represents, for it demonstrates as clearly as anything may the fact that Yahweh was not merely a symbol of tribal group interests. Furthermore, it illustrates an extraordinarily rare capacity for self-criticism (a capacity that becomes rarer under the impact of attitudes strongly promulgated as absolute truth by certain types of social scientists). The proclamation of the wrath of God is a very clear illustration of the ability on the part of ancient man to recognize a point of reference for individual and social behavior above and beyond the existing social interests and patterns of action

---

15:14–15 is of extraordinary significance ("You are my friends if you do what I command you. No longer do I call you servants [subjects = slaves], for the servant does not know what his master is doing; but I have called you friends . . ."). The background is Jer. 31:34: "No longer shall each man teach his neighbor, saying, 'Know the Lord,' for they shall all know me. . . ." This is also the context of Matt. 23:9: "Call no man father."

63. See below, chap. VIII, n. 3.

64. Ps. 32:9. Mitchell Dahood's emendations are largely unnecessary (*Psalms*, in *Anchor Bible* [Garden City, N.Y., 1966], 1:197), and actually obscure the meaning. The usual Hebrew idiom is *qrb ʾl* (e.g., Judg. 20:24 and cf. Assyrian *qarābu*), but the prepositions ʾel and ʿal are frequently interchanged.

65. *ABL* 1237:16. The Cimmerians are here called *zēr ḫalgatî* (*CAD*, Z, p. 87; cf. *AHW*, p. 313), which is very reminiscent of the *kalbu ḫalqu* of Amarna (*EA* 67) and ʾarammî ʾōvēd of Deut. 26:5. Cf. also the words of Nabal, below chap. V.

66. Deut. 32:26; Jer. 7.

and conviction. This is quite the opposite to myth, the specific purpose of which is to furnish metaphysical legitimacy for that which *is*. The radicality of the biblical tradition has been reduced to myth by theological and political establishments. But it is not the Bible that needs to be demythologized, it is theology itself, for the biblical faith cannot be reduced to mere forms of behavior or of theological propositions.

The covenant form is essential not only for understanding certain highly unusual features of the Old Testament faith, but also for understanding the existence of the community itself and the interrelatedness of the different aspects of early Israel's social culture. Here we reach a clear watershed, so to speak, in historical research. Do the people create a religion, or does the religion create a people? Historically, when we are dealing with the formative period of Moses and the Judges, there can be no doubt that the latter is correct, for the historical, linguistic, and archaeological evidence is too powerful to deny. Religion furnished the foundation for a unity far beyond anything that had existed before, and the covenant appears to have been the only conceivable instrument through which the unity was brought about and expressed. If the very heart and center of religion is "allegiance," which the Bible terms "love,"[67] religion and covenant become virtually identical. Out of this flows nearly the whole of those aspects of biblical faith that constitute impressive contrasts to the ancient paganism of the ancient Near Eastern world, in spite of increasingly massive evidence that the community of ancient Israel did not constitute a radical contrast to them either ethnically, in material culture, or in many patterns of thought and language. Indeed, early Israel is conceivable only within the framework of the cultural forms of the late Bronze Age, and its history and faith must be approached from the perspective of what preceded it, not from what it evolved into. The Old Testament Constantine, King David, represents a thoroughgoing reassimilation to Late Bronze Age religious ideas and structures. These readapted the authentic traditions of Israel just as radically as the later Achemenids readapted the original message of Zarathustra, or the period of Constantine readapted New Testament Christianity. All three cases are entirely analogous, illustrating (to put it as provocatively as possible) the dissolution of religion into politics. At the same time, the basis of solidarity

---

67. For a much later continuation of this semantic identification of subjective orientation with objective action, see Shakespeare's *King Henry VI*, Part III, act 3, sc. 2, where there is a play on the word involving two uses. Unfortunately, only the usage Edward was interested in seems to have survived in modern English language. Conversely, "hate" is the word for divorce in legal texts, and denotes severing of relationships in general, for which the best illustration is Ps. 139:21. Such "perfect hatred" is still the official policy of many religious denominations to the present day.

was no longer the covenant, but the myth of descent from a common ancestor.[68] Western man, who has done most of the thinking about the problem, unconsciously assumes that the political state is the social reality of utmost importance, and therefore sees the political establishment of religion as a step toward progress and "civilization." It is now quite clear that it is the decisive step back toward the age-old paganism of Babylonia and Egypt —the difference consisting merely in the forms of organization and the formal theological doctrines by which the establishment is justified. It is a sad irony, then, to have religion blamed for the consequences arising from the pagan constant of political realities.

Hence, the enormous emphasis upon cultural and social change characterizing much thought in the contemporary world is lacking in perspective. It is not very fashionable in intellectual circles to make such broad-gauge comparisons as those being presented here. However, it is necessary emphatically to point out that the rapid and large-scale change which we are undeniably witnessing today is not something radically new in the course of human history. On the other hand, the change in areas of human experience other than the technical—in individual and social ethic, in personal satisfactions in life, in human relationships, in short, in everything that involves persons as human beings— is hardly the sort that can be called "progress." To be sure, a carefully nurtured elaboration of religious ethical thought exists in all traditions, but most of the noble thought is abandoned when the social, political, or economic interests of persons or groups become involved. Unfortunately, there are very few issues of importance that do not involve such interests. I suggest that no understanding of either the ancient pagan world or the biblical tradition is possible unless one recognizes that the pagan gods and their symbols are most intimately involved with precisely such interests. If these are the only determinants of behavior recognized as legitimate and "normal" today, we have turned the clock back to the Late Bronze Age. Then perhaps we are in a position to understand both the radicality of the biblical tradition, and its stark realism in acknowledging the judgment and wrath of God as the inevitable results of such behavior. The conflict between ancient Israel and Canaan was neither a race war nor a class war, for they do not have such consequences as the biblical tradition. Rather, it was a revolution that equaled the agricultural (if such actually happened), philosophical, and industrial revolutions in importance. However, such a statement is defensible only from the point of view of a historical reconstruction of events, not from the usual theological attitudes

68. See chap. VII. Christianity had to reject this; Luke 3:8 indicates that already in the pre-Christian preaching of John the Baptist confidence in genealogical status is religiously nonfunctional.

toward biblical history, which reduce the events to mere ritual or social symbols. It is of course undeniable that the events *were* reduced to symbols in the course of biblical history; therefore historical reconstruction on the basis of all historical resources available to us must be undertaken for an adequate understanding of the early biblical religion itself.

The nature of the revolution may be summarized in the thesis that with Moses and early Israel, at least temporarily, a third stage of religious development occurred. The first was the economic-tribal religion in which the most important religious symbols and deities were projections of the economic base of existence of the tribe; this is prehistoric, but such primitive religion always survives. Functionally, it may be observed at work whenever any group with power regards its economic strength as a concern that takes precedence over all other potential obligations. The political religions of the ancient Near Eastern paganisms, in which the most important deities and myths were those that supposedly guaranteed the existence of the state, comprised the second stage. This seems the normal religion of "civilized" man, but it also is virtually prehistoric in origin. Here also the interests of the political power structure are predominant. Both of these stages become dysfunctional, first, because they inevitably threaten outside groups, and, second, because they eventually become so remote from the real lives and concerns of their own populations that their base of public support withers away. It was under these circumstances that the third phase, the biblical faith in which ethical concerns take precedence over all others, arose. Then, obligations were supposed to be binding even upon the state when it eventually emerged in imitation of the old paganism of pre-Mosaic times. The incompatibility of the state religion with the authentic religion of Israel is illustrated in the whole of biblical history, and it was not until the New Testament reformation that the attempt to combine the two was completely and systematically renounced, only to succumb to the temptation to become a political religion again under the Holy Roman Empire.

If the Kingdom of God seems hopelessly quaint and old-fashioned to contemporary society, let it be remembered that this was the only one of the religious traditions of the ancient world to survive in any influential way. If under the impact of population stress, affluence, and theologians more concerned with obtaining public attention than with comprehending their own subject matter, the tradition is again called into question, at least let it be said that there was a serious attempt to understand it in its own setting, which is not nineteenth-century imperialism or idealism, or seventeenth-century puritanism, or twentieth-century scientism. The essays following will, it is hoped, point out at least the fact that our understanding of biblical history, and therefore our misunderstanding of biblical religion, is grossly incomplete and incompetent. I submit that ignorance in this area is a luxury which mankind can no longer afford, for it leaves man with no basic hope other than to trust in violence: both the god and the fate of the Late Bronze Age.

THE SEQUENCE OF EVENTS

In order to place the following detailed studies of particular aspects of early Yahwistic faith into a historical context, and, also, to show the intimate relationship between historical events and the structure of the revolutionary movement, a brief resume of that history is presented here. It is not the only possible reconstruction of early biblical history, and subsequent archaeological work may well clarify the period further. Though any historical thesis must be compatible with the archaeological evidence, nevertheless, unwritten artifacts alone cannot produce a history. The biblical traditions need not be rejected, however. Rather, it is necessary to recover the original historical context of the events from the later forms—forms often chosen to be useful and "relevant" to a radically different and, indeed, often diametrically opposed sociopolitical situation.

The formative period of a religious community is characterized by numerous problems that are in many ways quite opposite to those of a traditional period. To bring unity to a large and varied population where hostility and intense competition existed before is an incredibly complex process. The political game of attempting to minimize internal stresses for the sake of preserving an always precarious unity is a much more familiar one. It should come as no surprise, therefore, if the minor successes and failures of the formative period are treated by tradition as though they were merely the usual political processes of resistance to and assertion of authority. As a matter of fact, the events of the formative period reflect a much deeper issue: whether there should or could be a unity which transcends the already existent complex of tribes, clans, kinship groups, and political power structures.

It is certain that over a period of several centuries prior to the Exodus, for many reasons, considerable numbers of persons from Palestine and Syria had become resident in Egypt. Quite a few were brought in by the Egyptians as prisoners of war[69] or hostages, and many were used as *corvée* labor in building projects.[70] Since some of these are termed ᶜApiru, it is highly

---

69. *ANET*, pp. 247, 261.

70. Moshe Greenberg, *The Ḫab / piru*, American Oriental Series, vol. 39 (New Haven, 1955), pp. 55–57, and Torgny Säve-Söderbergh, "The Apiru as Vintagers in Egypt," *Orientalia suecana*, 1 (1952): 5ff.

probable that they were exiled to Egypt precisely because they had been engaged in local activities deemed incompatible with the interests of the Egyptian Empire: political rebellion and dissidence.[71]

Egyptian policy concerning the resident aliens (who are called "Hebrews" in the biblical narrative) became more restrictive as Egypt itself came increasingly under attack from the so-called "sea-peoples."[72] Correspondingly, life grew more and more harsh for these people, especially under the *corvée* system. Though it is probable that each segment of the alien population retained something of its own religious, political, and tribal tradition or identity (particularly if the ancestor worship of the ancient world still prevailed among grass-roots populations of the time), the functional relationships of those old traditions would have been relatively limited under circumstances in Egypt. In order to furnish a unity in the diversity of populations from all over Western Asia, something new was needed.

It may well be that it all started with Moses' murder of an Egyptian overseer, occasioned by the brutality of the man. His attempt to put an end to a brawl between two ʿApiru brought the retort: "Who made you prince and judge over us?" "Are you threatening to murder me as you murdered the Egyptian?"[73] Thus asked point blank for his authority to exercise power, his resort to force and violence, Moses' response was flight to the desert no doubt motivated, as the biblical narrative indicates, by the fact that the murder had become common knowledge. Nevertheless, a question had been raised which is just as relevant today as it was in pharaonic Egypt: what is the ultimate authority for the exercise of force when the authorities themselves engage in clandestine murder? The great importance of this narrative for the origins of the religious ideology of ancient Israel is indicated by the fact that it is the only story preserved about Moses between infancy and his experience at the burning bush. Further, this passage is also paraphrased by Jesus when he is *asked* to exercise his power in behalf of "justice" (as seen by the *asker*) (Luke 12:14). The authority of the peacemaker must be based on something other than his superior ability to commit murder. Whether or not it is historically factual, it is an otherwise unmotivated element of the early biblical tradition that this experience of Moses immediately preceded his withdrawal into the desert. It must have been a traumatic experience—or have been seen as such by the religious community—to find

71. Cf. chap. VI. The same policy is known for Hittite, Assyrian, and Babylonian empires.

72. *ANET*, pp. 262–63. Cf. also the long discussion by C. F. A. Schaeffer in *Ugaritica* (Paris, 1968), vol. 5, pp. 638–760. See also chap. V; Exod. 1:8ff.

73. Exod. 2:14.

that the conflict between two Hebrews was identical to that between the Egyptian aggressor and the Hebrew victim. The only solution to both was superior force, to which the normal sequel was murder.

In spite of the several generations of biblical scholars who have assumed unquestioningly that early Israel consisted of primitive tribes moving slowly on the path to civilization, it is not possible to believe that biblical Israel's existence was based upon a superior ability to commit murder. On the contrary, early Israel rejected entirely the idea that God had delegated to some autocrat the legitimate power to put human beings to death. Early Israel rejected entirely the idea that God was merely the Ground of Being for some political monopoly of force. If a century before Moses the king of Byblos could write to the pharaoh: "Do you not know the land of Canaan, that it follows behind the one who has power?" then the expatriated Syro-Palestinians in Egypt included at least enough people to create a new community who could recognize the act of God in the deliverance *from* political power, which, as seems normally to be the case, was much more interested in its self-perpetuation than in the *quality* of its functioning for the benefit of *all* human beings subjected to its coercive force. It is, indeed, highly questionable whether or not political regimes regarded themselves bound (at least for long) to secure the maximum well-being of *all* their subjects in the ancient or modern world.[74] The best illustration comes from the son of Solomon whose statement proclaimed that the maximum exercise of force was the legitimate policy of the state.

It is within this historical context that the Sinai covenant becomes historically relevant. There is nothing in the Decalogue that constitutes any new or profound philosophical insight into ethics or law. It is not in the content, but in functional relationships that the Decalogue constitutes a revolutionary movement in human history. A covenant is not necessary to establish unity among a kin-bound group, particularly in Near Eastern history. On the other hand, a kinship-bound group is not likely to change the entire ideological basis of its religion, particularly when the "new God" had already been proclaimed to be the "god of your fathers." What happened at Sinai was the formation of a new unity where none had existed before, a "peace of God" among a "mixed multitude"[75] and tribally affiliated families who had in common only the deliverance from an intolerable political monopoly of force. Perhaps for the first time in history, a real

---

74. Most illustrative is Rehoboam's policy in 1 Kings 12. Ancient myths also seem to support the idea that the human masses exist for the purpose of supporting the politicoreligious establishment (*ANET*, Creation Epic VI, ll. 1–10 [p. 68]; ll. 110–20 [p. 69]).

75. Num. 11:4.

elevation to a new and unfamiliar ground in the formation of a community took place—a formation based on common obligations rather than common interests—on ethic, rather than on covetousness.

The community did not consist of a centralized power monopoly until two centuries later, by which time the constitutive factors which brought it into being were so diluted that they were probably very slightly understood, and actually rejected in favor of the more "normal" concerns such as "control of the environment," and the use of power to defend those concerns.

It seems probable that much of the population of Palestine and Transjordan were also subjected to political regimes which were alien in origin,[76] little concerned with the real well-being of those (mis-)governed, and constantly engaged in attempts to extend their own military power at the expense of their neighbors.[77] The message of a new element in human history, associated with the miraculous escape from Egyptian control, and reinforced by the defeat of attacks by Sihon and Og (both almost certainly newcomers to political and military power in Transjordan) must have been extremely attractive to the people of the region. It is certain that the entire area of Palestine and Syria was either periodically or constantly subjected to violence and disruption because of the political wars and struggles for control on the part of local dynasts and politically ambitious petty kings, plus the sporadic efforts of the Egyptian Empire to maintain control and those of the Hittite Empire to extend its control. If it is justified at all to speak of the ideological and ethical bankruptcy of politics, the period with which we are dealing would seem to be a parade example.

We know that most excavated cities of Palestine suffered violent destruction during the second half of the thirteenth century B.C.,[78] but whether or not the biblical narratives of "conquest" can be correlated with the archaeological evidence as easily as many scholars assume is highly questionable.[79] In the first place, such destruction levels are virtually universal at this time throughout the whole of the eastern Mediterranean area, which strongly suggests that some much more widespread historical process is involved than the mere invasion of a small band of supposedly "primitive

76. See chap. V. This is particularly true of the north, where both the names and the titles of kings are of Syro-Hittite origin.

77. The old poem of Num. 21:27–30 certainly commemorates a pre-Israelite aggressive war of such nature.

78. A good survey is given by P. Lapp, "The Conquest of Palestine in the Light of Archaeology," *CTM* 38 (1967): 283–300. Little or none of the archaeological evidence he cites is incompatible with the thesis presented here. On the other hand, his equation of new pottery style with an alleged new wave of invasion is historically naive.

79. Wright, *Biblical Archaeology*, pp. 69–84, for the classical archaeological interpretation. Cf. also *LIS*, pp. 124–39, for the criticism of this position by the German school especially of Alt and Noth.

semi-nomads."[80] Secondly, if it could be proven that the famous Merneptah stele which mentions Israel has to do with purely internal or border problems rather than with a large-scale military expedition to remote areas,[81] then it would follow that a social group called Israel was in or near the borders of Egypt about 1225 B.C. This would make Merneptah the pharaoh of the Exodus, and the "conquest" a phenomenon of a generation later, after about 1200 B.C.

The fact that new composite "tribes" are a common phenomenon in anthropological and historical data *following* some historical or even natural catastrophe would correlate very well with this.[82] The destruction weakens the older social and political structures to such an extent that a new and more broadly based social organization can arise. The destruction levels revealed by archaeology in Palestine would have been caused not by the Israelites, but rather are part of the common experience of the population that made vivid the desirability and need for a new type of community. This could bring about the peace and secure a new cooperation for rebuilding a shattered society and economy.[83]

If we should characterize the old kinglets (largely non-Semitic in origin, as we know from Amarna) as semiprimitive chiefdoms (still another link in the chain of primitive societies between the food-gathering band and the civilized state) in which the chief is the redistributor of wealth,[84] then there is a curiously ironic twist in political evolution at the transition from the Late Bronze to the Early Iron Age. If the king (i.e., the state) owned the means of production in the Late Bronze Age city-state—in other words a form of primitive communism—the sequel was the withering away of the state under the old Israelite federation, and the introduction of a truly egalitarian society under the Kingdom of God. Pointing toward the existence of such a society is the still controversial fact that Early Iron Age housing is not characterized by the great contrast between luxurious villa and miserable hovel that has so frequently been observed in Late Bronze Age archaeological levels.[85] Further, its existence is most abundantly illustrated in the content of the Covenant Code (Exod. 21–23), which is characterized by a rigorous

80. Similarly, the idea of a natural catastrophe (Schaeffer, *Stratigraphie comparée et chronologie de l'Asie occidentale* [London, 1948], and *Ugaritica*, vol. 5, p. 760–68) is historically naive, and does not account for the political and cultural changes that ensue.

81. Étienne Drioton, "La Date de l'Exode," *RHPhR* 35 (1955): 44; *LIS*, p. 65, no. 3.

82. *PSO*, pp. 119, 136–37.

83. The attitude toward the old state system is well illustrated in Judg. 9, the parable of Jotham.

84. *PSO*, pp. 143–44.

85. Cf. Albright on the evidence from excavations, *FSAC*, p. 285.

concern for equity and reciprocity,[86] and a complete lack of any contrast between the nobility and the commoner. It seems quite clear the nobility of the Late Bronze Age is usually the military aristocracy and the civil bureaucracy (whose economic affluence is due to their proximity to the source of power), and both constitute the priesthood as well.[87]

If the major functions of religion in the Late Bronze Age were the guaranty of political legitimacy of the state, and the correlative guaranty of economic abundance, why the entire process constitutes a religious revolution is then apparent. Something had to take precedence over and destroy the religious ideology, the myth, upon which political legitimacy rested. Similarly, a concept of the process of productivity had to emerge that would liberate the population from the age-old tutelage of the great fertility goddess, and the rituals associated with her.

Both of these functions necessary to the revolution were met within the form and content of the Sinai covenant. If the petty kings themselves recognized that they were subject to the gods,[88] then it is a simple step to recognize that God can rule without the kings—in fact, the gods of the kings are mere projections of the sovereignty itself, functional politically, but subject to ethical judgment on the grounds of gross misuse of human beings and insensitivity.[89] If this be the ground for rejection of the royal sovereignties,[90] then those who appealed to Yahweh as a justification could not be characterized by the same sort of callous indifference to humanity. The Sinai covenant and its successors established this ethical obligation to the real needs of persons and society as the absolute upon which life and death depended. At the same time, in the curses and blessings formulas, the unpredictable and uncontrollable productivity of fields and flocks was made a function of that ethical obedience—following one pattern, already present in pre-Mosaic thought, that is well attested in Hittite treaties.[91]

The biblical revolution, then, saw, a radical redefinition of what we now term religion. Under the old primitive tribalism, religion consisted mainly of elaborate rituals performed, first, for maintaining the solidarity of the group and its complex of authority structures,[92] and, secondly, for

86. To say nothing of the equally rigorous concern for the protection of aliens and slaves (Exod. 21:5-6, 20-21, 26-27; 22:21-24).

87. *AT*, p. 39. S. Smith, "A Preliminary Account of the Tablets from Atchana," *AJ* 19 (1939): 43-44. It does not prove, of course, that all *mariannu* everywhere were priests, but, on the other hand, *AT* 15 can hardly constitute a unique case.

88. Cf. chap. III—and the very fact that covenants constantly were oaths watched over and sanctioned by the gods.

89. Cf. Ps. 82, and the discussion in chap. III.

90. Josh. 12.

91. Delbert R. Hillers, *Treaty Curses and the Old Testament Prophets* (Rome, 1964), *passim*.

92. *PSO*, p. 126.

influencing the supernatural world to give the group what it desired, and protect it from evils it could not control. Over against this, the new concept of religion consisted of man's voluntary submission to the will of God defined in ethical terms that were binding beyond any social or territorial boundary. The new religion actually destroyed most of the traditional boundaries over which so many battles had been fought from the time of the Amarna age. If the center of the old paganism was concern for perpetuating the king's control over all his enemies, the new proclaimed that no one but God was, or could be, in control. His authority was exercised in the first place by the community's obedience to His commands, and secondly by His control over all these powers of nature and history that man individually and corporately could neither control nor predict, but the functioning of which was also subjected to ethical interpretation in the covenant formula of curses and blessings.

Any history of the origins of ancient Israel must start with, or at least account for, the sudden appearance of a large community in Palestine and Transjordan only a generation after the small group escaped from Egypt under the leadership of Moses. At the same time, it must account for the fact that from the earliest period there is a radical contrast between the religious ideology of Israel and those of the preceding periods and neighboring groups. In spite of that contrast, virtually all specific *formal* elements in early Israelite culture and ideology have impressive analogues in pre-Israelite or other foreign sources.[93]

All of these historical problems are most easily explained by the fact that the religious community was based upon a covenant, which is most strongly emphasized in biblical tradition. The enormous and rapid growth in size can only have been possible through the entry of existing population groups into the religious community through covenant, which is actually the opposite to law, and which functioned in a way entirely analogous to a modern constitution and bylaws. Biblical tradition actually gives narratives or indications of three such historical covenants, in which radically different social contexts are either demonstrable or necessarily inferred. The first, at Sinai, was the instrument by which the "mixed multitude" became vassals (servants) of Yahweh after the miraculous escape from Egypt. The second covenant, the last significant act of Moses' life, was radically transformed by much later tradition into the "second giving of the Law"—Deuteronomy.[94] In the narrative, as in the case of Sinai, the solemn acceptance of obligation

93. Of course, many of them are mere formal continuities from the old pre-Yahwistic "Canaanite" and Anatolian cultures which characterized the Palestinian scene prior to the creation of the socioreligious unification.

94. Deut. 29:1–31:13 (Heb. 28:69–31:13). It has often been observed that the entire book of Deuteronomy has precisely the structure of the Late Bronze Age covenant form (see especially McCarthy, *Treaty and Covenant*).

is at least represented as having taken place *after* the defeat of Sihon and Og. The Transjordanian population segments joined the covenant community and were confirmed in their enjoyment of lands and fields, free from the claims of the northern kings, but with the condition that they were bound to serve upon Yahweh's command.[95] Eventually, the Transjordanian segments were organized into the well-known "tribes" of Reuben, Gad, and half-Manasseh, but, as is well known, the tribal structure of Transjordan was unusually unstable.[96] There is nothing surprising about this—it is a constant in tribal organization all over the world and all through history.[97] Large social organizations are normally ephemeral, but the disappearance of a named tribe has nothing to do with the continuity of lineages, persons, or even cultures, unless other factors are involved. The old gives way to and is absorbed by the new (as anyone involved in a modern church merger knows). Furthermore, when social organization changes, the genealogy is brought up to date to reflect the new social situation, and all segments receive a common ancestor.[98]

The third covenant is much better attested (in Joshua 24) than the second, and has already been adequately discussed.[99] Here, I need only emphasize that according to the biblical narrative it took place *after* the successful completion of the wars by which the various power structures of the land were either destroyed or reduced to virtual impotence. The most vivid description of what happened is found not so much in the dramatic events that were celebrated in song and story as in the little-noticed chapter 12 of the Book of Joshua where we find the list of thirty-one kings whose holdings were allotted to the various populations subsequently organized into the system of twelve tribes. It is a most vivid picture of the incredibly complex political fragmentation of the land prior to the revolutionary unification.

The greatly enlarged tribal organization thus became socially functional, each group under its *naśîʾ*, but we are extremely ill informed as to the range of social functions of either the tribe or the tribal head. Most probably, the functions were largely military and ceremonial.[100] Indeed, the purpose of the revolution was the creation of a condition of peace in which every man

95. Deut. 3:18–22. See S. Herbert Bess, "Systems of Land Tenure in Ancient Israel" (Ph.D. diss., University of Michigan, 1963), pp. 97–117.
96. Martin Noth, *The History of Israel* (2d ed.; New York, 1960), pp. 155–62.
97. *PSO*, p. 114.
98. Chap. VII.
99. Mendenhall, "Covenant Forms"; Hillers, *Covenant*, pp. 58–71.
100. Compare *PSO*, p. 114, Num. 1 (cf. G. E. Mendenhall, "Census Lists," *JBL* 77 [1958]: 52–66) for military functions, and Num. 7 for ritual involvement. The priestly code of course regards virtually all ritual as the private prerogative of the priests and Levites—which is almost certainly the result first of increasing specialization of functions during the Monarchy, but probably not carried to its logical conclusion until the archais-

could sit under his own fig tree and under his own grapevine,[101] doing "what was right in his own eyes"[102]—a description of self-determination and freedom from interference or harassment by the king's bureaucrats or military aristocracy. In fact, the latter phrase is used in the Late Bronze Age to describe the freedom of choice and action that was then regarded as a proper prerogative only of an independent "great king," over against the status of his vassals, who were not thus free agents.[103]

Furthermore, individuals probably did not characteristically identify themselves with the tribe to which their village was assigned.[104] Our evidence of this is the fact that the names of the Twelve Tribes occur very rarely in narratives as gentilic designations either of villages, lineages, or persons.[105] The lineage or the village was the unit of primary importance, and the tribe was a superstructure with very limited social function. The gentilic "Israelite" does not even occur in biblical Hebrew (except, according to scholarly consensus, once as a textual corruption, and twice in a very late passage, once in feminine form).[106] There was no such thing as an ethnic group of "Israelites" in this early period. Throughout history, a feeling of ethnic identity has been the product of a very long and complex process of continuity and contiguity.[107] The basis of solidarity changed as the nature of the social organization changed. What was at first a religious and ethical unity created from a very diverse population that did not even speak the same dialect of West Semitic[108] gave way to an uneasy political unity that

---

ing return to old traditions after the Restoration. See now also the article by Milgrom cited in chap. V, n. 62.

101. 1 Kings 4:24—where it describes the security and peace of the reign of Solomon! Cf. Mic. 4:4, where the phrase describes the future reign of Yahweh over *all* the earth with peace and security, based upon the Word (= Torah) of the Lord. It is generally believed to be a late poem.

102. Judg. 17:6; 21:25. Cf. Deut. 12:8, and Prov. 12:15.

103. Akkadian: *ippuš kima libbišu* 'acts in accordance with his own desires.' As in the Bible, it is used only as a reproach for evil deeds (except Deut. 12:8 where it certainly refers to a tradition of wide variety in cultic practice that is excused temporarily), *EA* 88:9–11; 104:17–24; 108: 8–13; 109:9–11; 116:66–71; 123:38–40; 139:11–12; 140:9–10.

104. G. E. Mendenhall, "The Relation of the Individual to Political Society in Ancient Israel," in *Biblical Studies in Memory of H. C. Alleman*, ed. J. M. Myers (Locust Valley, N.Y., 1960), pp. 92–93.

105. *Ibid.*

106. Lev. 24:10. Cf. 2 Sam. 17:25, and also 1 Chron. 2:17 which is certainly correct.

107. *PSO*, pp. 106–7. According to this theory, ancient Israel would necessarily be termed the "Empire of Yahweh." The theories of Service are far too simple to be useful historically, although the pattern is the same.

108. As the *šibbolet* incident proves, regardless of what linguistic explanation may be preferred.

soon divided into two. Not until the political and religious institutionalism was destroyed did the basis again shift to a concept of ethnic unity, long after the Babylonian Exile,[109] and that was thoroughly rejected and challenged by the reform movement that we now call Christianity.[110]

Israel was the name of the large social organization that constituted the population ruled by Yahweh,[111] and, as the prophets all pointed out,[112] when it ceased so to be ruled it ceased to have legitimate grounds for corporate existence. The much praised concern of the prophets for "social justice" stems from the fact that the transition from a religious community to a political power structure had eroded away the old religious ethic in favor of "being like all the rest of the nations," obsessed with power, concentration of wealth, and competition in the insane world of power politics.[113]

The radical difference between persons and political power structures is just as basic to biblical thought as it has been throughout history.[114] The biblical revolution was a marvelously creative answer to the old dilemma of loyalty to the small, primary community versus loyalty to the large community necessary to a rising population with its more complex economic and political problems. It survived long enough to prove that it could function effectively, but it seems to have become dominant only *after* a major catastrophe: the Babylonian Exile and the destruction of Jerusalem by the Romans. One would think that a major social catastrophe ought not to be necessary to emphasize the fact that persons and their well-being should take precedence over abstract political structures and social organizations. The political structure exists for the well-being of those who live under its control, not the opposite.

In summary, then, early Israel was the dominion of Yahweh, consisting

---

109. Neh. 13:3, 23–29.

110. Rom. 10:12: "For there is no distinction between Jew and Greek; the same Lord is Lord of all and bestows his riches upon all who call upon him." See Gal. 3:28. The ethnic and cultural differences are just as irrelevant to the rule of God as are those of legal status (under the pagan secular law) or of sex. This is nothing new; Isa. 49:6 and many other passages emphasized as strongly as possible that the salvation and "teaching" of Yahweh should extend to all nations (Matt. 12:50; Luke 13:29).

111. The name is certainly simultaneous with the formation of the community in Egypt, perhaps even a pre-Israelite name, and continues to be used as the designation of the *religious* community in the New Testament, "the Israel of God" (Gal. 6:16). Cf. Ugar. *yśr'il*, *PTU*, p. 146.

112. I.e., the pre-Exilic prophets; only Zephaniah and Nahum lack a prophecy of destruction of the state. The former is almost certainly to be assigned to the period of Josiah's reform movement, 623–609 B.C., and the latter is a highly specialized curse upon Nineveh.

113. Isaiah particularly rings the changes on these motifs.

114. See Mendenhall, "Relation of the Individual."

of all those diverse lineages, clans, individuals, and other social segments that, under the covenant, had accepted the rule of Yahweh and simultaneously had rejected the domination of the various local kings and their tutelary deities—the baalīm. As a necessary corollary, Yahweh was the one who exclusively exercised the classic functions of the king, as described in the prologue to the Code of Hammurabi and in other early codes as well. In the earlier code of Lipit-Ishtar, we read: "Lipit-Ishtar, the wise shepherd whose name had been pronounced by Nunamnir—to the princeship of the land in order to establish justice in the land, to banish complaints, to turn back enmity and rebellion by the force of arms, [and] to bring well-being to the Sumerians and Akkadians . . . ."[115] The administration of law internally, the waging of war, and the economic well-being of the diverse population are here already the three prime functions of the king, and they correspond to the modern functions of a state as a monopoly of force.

In early Israel, all these functions were directly attributed to Yahweh, whose function as a leader in war has frequently been an embarrassment to a liberal theology unaware of the historical context of the whole ideology. The creation of a state of war could legitimately come about only through Yahweh Himself, but the declaration was released, so to speak, through a spokesman—prophets (or prophetesses as the case of Deborah indicates), "men of god," or the assembly of the representatives of the tribes. The second function of Yahweh as judge and lawgiver receives much greater emphasis, though, as in any other kingship, the daily routine of court cases was delegated to and operated through the village courts, but was thoroughly subjected to the royal policy of God, who remained the final court of appeal: ". . . he will cry out [i.e., appeal] to me and I will hear, for I am compassionate."[116]

The third function of general economic welfare operated primarily through the curses and blessings formulas, which have to do largely with the natural world, but with the external pagan world as well, since economic well-being was hardly compatible with a state of constant warfare. All of these constant elements in human political society are thus subjected to ethical controls defined as the rule of God, and are given an ultimately ethical (and therefore human and historical) interpretation as opposed to the metaphysical interpretations of ancient paganism. To reject the ethical control, or to accept other grounds for behavior and choices, such as the ethically neutral technical means of divination or necromancy or even "wisdom," was to reject God Himself, and was punishable by death in the

115. *ANET*, p. 159.
116. Exod. 22:27. In biblical faith this is certainly an important origin, if not *the* origin, of prayer.

early law code.[117] The ethic had predictive value—in fact, the ethic consisted in large part of the ability of persons to predict the consequences of their behavior as they affected other persons and caused them to react. This is one of the foundations of pre-Exilic prophecy, which enabled them to predict the course of history with uncanny accuracy.[118]

The prophetic message is relatively simple: the rejection of those ethical controls that were identified with the rule of God constituted the rejection of God Himself, and therefore the corporate existence brought about by that divine rule could no longer be legitimized by appeal to Yahweh. On the contrary, it must be destroyed as the enemy of Yahweh. But there is nothing in the prophetic message that would prescribe the death penalty for all persons:[119] rather it is the social institutions of state and temple that must go, once they have ceased to be responsible either to Yahweh or to those in society who most need protection.

As a corollary to this rule of God: if a military leader, judge, or even farmer were blessed with success or outstanding achievement in his occupation, to claim glory or social prestige on such grounds would constitute an act of rebellion or treason.[120] Joab recognized this principle at the siege of Rabbath-Ammon,[121] as well he should have in view of the enormous hardship and danger he shared with David because of the empty-headed adulation of the successful military leader by the young women of the country.[122] This principle of agency is essential to any functioning political structure, but it operated in a fashion radically different from that of the age-old paganism in which the military leader is the great benefactor who is thereby entitled to the gratitude and obedience of the populace. The receipt of a benefit always places the recipient under obligation, but in the rule of God, no person whose activity resulted in benefit to others could claim their obligation to himself, simply because he himself was acting as an agent of superior command.[123] With the growth of practical atheism, however, nothing existed as the

117. Exod. 22:18 very probably has to do with necromancy; cf. the witch of Endor narrative (1 Sam. 28:9).

118. See the study of Hillers, *Covenant*. See also his *Treaty Curses*, pp. 80–89.

119. The "remnant" is a recurrent theme from Amos to Jeremiah.

120. Gideon's words are an authentic expression of the old ideology (Judg. 8:22–23).

121. 2 Sam. 12:26–28.

122. 1 Sam. 18:7–8. The same is true of the prophet. Though modern thought regards it as incredible presumption to present a message as the authentic "Word of the Lord," as do the prophets, ancient thought regarded it as even more arrogant and presumptuous to present such messages as the prophet's own idea. This is well recognized in the New Testament period: "He who speaks on his own authority seeks his own glory" (John 7:18).

123. Luke 14:12–14. Cf. also Luke 17:7–10.

superior of the military general, and the ancient story of man's subjection by man ended with the mad policy of Rehoboam. Politics has made little progress since, and the process of political revolution is usually based upon the blind devotion of masses in hope of gaining wealth by following a leader who promises everything.

II

THE MASK OF YAHWEH

The doctrine that the king is immortal and invisible seems at first glance to be the kind of idea that would belong to the archaic "primitive cultures" of the ancient Near East. It may come as somewhat of a surprise to learn that this English doctrine of kingship was well developed by the time of Queen Elizabeth I, and was not really challenged until late in the nineteenth century, hardly a hundred years ago.[1] As Kantorowicz states, "The king is immortal because legally he can never die . . . ." Moreover, the king is invisible, and, though he may never judge despite being the "Fountain of Justice," he yet has legal ubiquity: "His Majesty in the eyes of the law is always present in all his courts, though he cannot personally distribute justice."[2] Similarly, the king has a "soul," his immortal aspect, that migrates from one incarnation to another.[3]

Thus we find in quite recent times a theory of kingship which Kantorowicz aptly describes as "the king's two bodies." The usefulness of the concept is illustrated by the English Parliament's having regarded itself as acting within the law when it exiled King Charles I to Oxford. His haloed body remained in Parliament,[4] for it was only the unhaloed natural body of Charles that was exiled. The whole concept is of course derived from early church theories about the nature of Christ, which were soon elaborated so far as to conclude that even the ass on which Jesus rode on Palm Sunday had two natures, the haloed eternal nature that perpetually served the divine king, and the unhaloed nature that had to be returned to its owner, since Jesus had only borrowed it. To return to the fate of Charles I, Kantorowicz wittily notes that it would be unjustified to conclude that the unhaloed ass is always

Note: Unless otherwise noted, the photographs in this chapter were taken by the author.

1. Ernst H. Kantorowicz, *The King's Two Bodies* (Princeton, N.J., 1967), p. 3. I owe to my colleague Professor Stephen Tonsor the reference to this most valuable work. It is he who first saw its relevance to the complex problems of ancient concepts of sovereignty.

2. *Ibid.*, pp. 4, 13.

3. P. 13.

4. P. 86.

to be found at Oxford, or that the haloed ass is always to be found in Parliament.[5]

It is quite clear, however, that the medieval theology of the two natures was in turn derived from political ideology of the Roman Empire. The emperor Caligula, for example, erected a temple to his own *numen*, ordained priests to officiate at cultic rites dedicated to his image, and daily had the idol clothed with an exact duplicate of the clothing he himself wore on that day.[6] The Roman emperor always had a *genius* that was regarded as divine and an object of worship, but it was frequently identified with the *genius* of some god, usually Jupiter or Hercules.[7]

It is generally admitted that such conceptualizations reached the West as a consequence of the empire of Alexander the Great.[8] At the Ptolemaic temple of Horus at Edfu in Upper Egypt, not far from the famous Aswan Dam, an elaborate narrative of the battles between Horus and Seth has been found. In many ways, this seems to serve as a bridge between ancient and recent, and between East and West.[9]

In the myth, Horus is traveling in the solar boat with Rē-Ḥarakhte when they see the cohorts of Seth ahead. When Horus asks for and is granted permission to attack by Rēᶜ, he ascends into the heavens as a great winged disk (great flyer) and throws the enemy into such confusion that they can neither see nor hear. They proceed to attack each other until none of them are left.[10] At present the narrative cannot be interpreted historically, although various specialists have suggested that it is a disguised reference to some historical conflict.[11] However, the attempt to find some historical symbolism in the narrative is much less important than understanding its conclusion:

> Then shall the king himself say: "I am the god's avenger [?] who came forth from Beḥdet, and Horus of Beḥdet is my name" [four times]. Let this utterance be recited when trouble occurs, and the king shall not be afraid, but his foes will be slain before him, and his heart will rejoice over them immediately, and [each] one will slay his fellow immediately, as befell the enemies of Rē-Ḥarakhte when Horus of Beḥdet [flew] against them as the great Winged Disk. This image shall be made with the face of the king to this day.[12]

5. *Ibid.*

6. Pp. 501, 502. Suetonius *Caligula* 22.3. Dio Cassius LIX, 28.5.

7. Kantorowicz, *King's Two Bodies*, pp. 502–3.

8. P. 497. Cf. also Erwin R. Goodenough, "The Political Philosophy of Hellenistic Kingship," *Yale Classical Studies*, 1 (1928): 55–102.

9. For a translation see A. M. Blackman, and H. W. Fairman, "The Myth of Horus at Edfu," *JEA* 21 (1935): 26–36; 28 (1942): 32–38; 29 (1943): 2–36; 30 (1944): 5–22.

10. H. W. Fairman, "The myth of Horus at Edfu—I," *JEA* 21 (1935): 28.

11. *Ibid.*, n. 2: ". . . there seems to be a certain substratum of historical fact on which they are based."

12. *Ibid.*, p. 36.

The myth of the winged sun disk informs us, first, that the great god Horus, Lord of Heaven (= Baᶜal šamēm), in action is the winged sun disk, and second, that the reigning Ptolemy is Horus. One could almost justifiably conclude that the "soul" of the king is the "body" (i.e., the external manifestation of the god), and the two meet in the symbol of the winged sun disk, which represented both and probably had for at least two and a half thousand years before the temple of Horus at Edfu was built. Though Gardiner maintains that the myth of the winged disk was merely local and provincial,[13] it can now be stated with confidence that certain of its motifs were current from India to Greece in the Iron Age, and lie at the foundations of the earliest biblical traditions. To the examination of that tradition we now turn.

Gardiner says of the meaning of the winged sun disk that, "the evidence thus all goes to show that Winged Disk and name of king are so inextricably interconnected and blended that we cannot but regard the symbol as an image of the king himself, though simultaneously also of Rēᶜ and of Horus, all three united into a trinity of solar and kingly dominion."[14] To date, the winged disk in Egyptian art has been found only from the Fifth Dynasty on, though a pair of wings over a falcon which certainly represents the king is seen on a comb of the First Dynasty King Djet. Only in the reign of King Sahure of the Fifth Dynasty does the pair of wings with solar disk appear, and, still later, we find the pair of cobras depicted (fig. 1).[15] Especially since my investigation began with Canaan and the Syro-Hittite region, I am most happy to concur with Gardiner's conclusions concerning the origin of the winged disk:

> Thus in this forerunner of the Winged Disk symbol [i.e., the wings and falcon of the Djet comb] we discern the fusion of the sun-god Rēᶜ, of the falcon Horus, and last but not least, of the reigning king whose name fills the universe, and whose protection extends over both North and South; just as the sun, in the guise of a flying falcon, spreads light and colour like wings over the entire land. . . .[16]

To say, then, that the winged disk represents heaven is already a gross oversimplification even for Old Kingdom Egypt.[17]

13. Alan Gardiner, "Horus the Behdetite," *JEA* 30 (1944): 23–60.

14. *Ibid.*, p. 51.

15. *Ibid.* Cf. also W. S. Smith, *A History of Egyptian Sculpture and Painting in the Old Kingdom* (London, 1946), p. 324; and J. H. Breasted, *The Dawn of Conscience* (New York, 1933), p. 57 and fig. 6, where the capstone of the pyramid of Amenemhet III is already ritually identified with the face of the deceased king. The eyes of the king are surmounted by the winged sun disk.

16. Gardiner, "Horus the Behdetite," p. 49.

17. *CS*, pp. 205–15. It should be pointed out, however, that this work was published prior to the studies of Gardiner and Smith, cited in notes 15 and 16.

Fig. 1 Offering stela of Amenhotep III. Karnak, outside Hypostyle Temple. Early fourteenth century B.C.

It seems necessary to conclude that Mesopotamian tradition is strikingly similar as far as the early stages are concerned, though the associations of the various symbols vary. In earliest historic times we find the eagle with out-stretched wings which can be identified with the Im-Dugud bird of Ningirsu, the warrior god of Lagash[18] (fig. 2). In Mesopotamia, this is not so much associated with the sky as with a thunderstorm. The sun symbol is found even

18. T. Jacobsen, *JNES* 12 (1953): 167, n. 27.

Fig. 2 Im-dugud bird with stags. From Lagash. British Museum No. 114308. Photograph courtesy of British Museum.

in remote prehistoric times,[19] but its specific associations are a matter of speculation that we need not enter into here. At present it seems quite clear that the winged sun disk did not enter into Mesopotamian iconography until well after it had become established in Syrian and Canaanite iconography, probably under Egyptian influence, as Frankfort and the general consensus hold.[20] However, the way was so well prepared for the sun disk in Mesopotamia, Syria, and Anatolia, both in the artistic motifs very common even in proto-literate times and in conceptualizations, that we would have relatively little difficulty in accounting for the origin of the winged disk were there no prior Egyptian evidence.[21] The winged disk in American Indian art seems to be an independent invention, and any spectacular sunrise or sunset could easily

19. H. R. Hays, *In the Beginnings* (New York, 1963), pp. 138–39. For the sun symbol in wheel form, however, cf. H. Z. Koşay, *Alaca Höyük Kazısı* (Ankara, 1951), plate CXCIV.

20. *CS*, pp. 208–9.

21. For the spread wings, compare the Im-Dugud Bird in text above citation of n. 18, and the sun symbol, n. 19. Cf. also the *šurinnu* and discussion by E. D. van Buren, *Symbols of the Gods in Mesopotamian Art* (Rome, 1945), pp. 90–91.

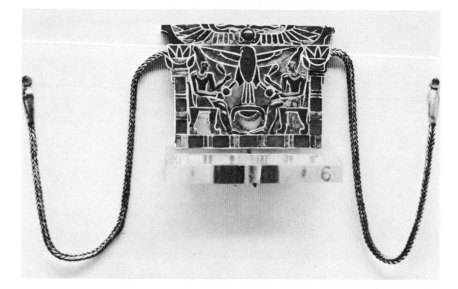

Fig. 3 Pectoral of Yapa-šemu-abi, king of Byblos. Mid-eighteenth century B.C. National Museum of Beirut. Photograph courtesy of Dr. Roger Saida.

have inspired it in any culture where the clouds are regarded as analogous to birds.

If Gardiner is correct in seeing the winged disk as a symbol that combined the sky bird Horus with the solar Rēᶜ,[22] we find an even more specific association in eastern Syria almost exactly at the time the winged disk began to appear in Syria. Išme-Dagan writes to his "brother" as follows: "Your throne is truly your throne; I hold in my hand the gods Šamaš and Adad."[23] It is most tempting to see this as a reference to the fact that the successor to Šamši-Adad I of Assyria had two major seats of power which are frequently referred to in the Mari archives: Šubat-Šamaš, 'dwelling of the sungod,' and Šubat-Enlil, 'dwelling of the storm god' (the latter of course being identified with West Semitic Hadad). The gods symbolize the continuity of the sovereignty that is now in the control of Išme-Dagan. However, as far as I have been able to determine, it was not until the Mitanni period a couple of centuries later that the winged disk penetrated into eastern Syria and thence to northern Mesopotamia, most probably from the Canaanite area where it already occurs in the mid-eighteenth century B.C.[24] (fig. 3).

22. Cf. n. 16.
23. (*ARM*)T IV 20:14.
24. "The earliest indubitable rendering of the winged disk on an Asiatic monument, the sealing of King Shaushattar of Mitanni . . ." (*CS*, p. 209, see plate XLII a, b,

Fig. 4 Seal impression no. 67. From the collection of the Department of Near Eastern Languages and Literatures, University of Michigan.

In Syrian and Mitanni art, however, and subsequently in Hittite art, which certainly depends upon Syria, something new was added. The winged disk most characteristically rests upon a standard, itself a much older motif in Mesopotamia.[25] An UR III seal impression in the University of Michigan collection shows a spread-eagled bird, probably Im-Dugud, in a pose almost precisely anticipating the winged disk of Mitanni (fig. 4). The pole has been interpreted by Frankfort as the "maypole," symbol of fertility.[26] The Djed

---

e, and o). This must be modified by the findings of Edith Porada, who dates the disk to "within two or three generations following the time of Hammurabi of Babylon" ("Syrian Seal Impressions on Tablets Dated in the Time of Hammurabi and Samsu-Iluna," *JNES* 16 (1957): 195, plate XXXI, 7). This is of course presumably the earliest possible date for the seal, which at any rate would credibly make it earlier than the sealing of Shaushattar, regardless of the dating used. Frankfort evidently overlooked the pectoral of Yapa-shemu-abi, or did not regard it as an "Asiatic monument" (*CS*). The Syrian style discussed by Porada is an extremely likely candidate for the medium that transmitted the winged disk from the Egyptianizing coast to the eastern Syrian region of Mitanni art, and to Anatolia as well, though it is most tempting to connect the diffusion with the Hyksos regime in Egypt.

25. *CS*, pp. 264, 275.

26. *Ibid.*, p. 275. Dr. Porada thinks that the University of Michigan impression no. 67 of unknown origin is not genuine, but the motif is very well attested in UR III seals (cf. van Buren, *Symbols of the Gods*, p. 90, for the column surmounted by a disk, which need not specifically be a symbol of the god Shamash).

pillar of Egypt has also been compared,[27] to note the parallelism of motifs in the two areas. With the passage of time, the sacred pillar sprouted leaves and bore fruit (figs. 5, 6). This "tree of life" motif cannot then be separated from the mother-goddess, the fertility figure, called Asherah in Old Testament tradition.[28] By the time of the Late Bronze Age, therefore, we find both the conceptualization and realization in art corresponding to that found in the Old Testament tradition. Besides the pillar, however, the winged disk is very frequently supported by two pillars or two winged bull-men[29] (fig. 8). These creatures are of course inseparable from the vehicle of deity as may be seen in statements ranging from the tenth-century Psalm 18 to early India: "He rides on the cherub and flies, / and soars on the wings of the wind."[30] As Zimmer describes the Hindu counterpart: "Ablaze with the heat of the glowing sun, drying up the moisture of the land, the 'fair-feathered,' golden-winged, griffin-like master of the sky violently attacks.... His proper name is Garuda ... represented, generally, with wings, human arms, vulture legs, and a curved, beaklike nose ...."[31] Garuda is the vehicle of the supreme god, Vishnu. The dependence of early Hindu myth upon the ancient Near East has already been pointed out by Marvin Pope[32] and Fontenrose.[33] The description provides us with the four essentials: the sun, the golden wings, the

27. *CS*, pp. 206–7.

28. This is not, of course, to argue for any specific linear evolution, but merely to observe that the various artistic and religious traditions must be interrelated in some complex way we shall probably never be able to describe historically. Ancient thought is associational, not "scientific," and therefore tends to create the maximum of relationships between experience, language, and art, not the minimum which is so characteristic of modern overspecialization. For the tree goddess, compare the ivory inlay tree with winged disk, which alternates with the nude female with winged disk (fig. 7), at Nimrud (M. E. L. Mallowan, *Nimrud and Its Remains* [London, 1966], 2: 458, 549, and 373, 481; 1: 43, 230). Note also the Greco-Sicilio-Carthaginian goddess Malophoros whose cult in Sicily was recently explored by D. White, "The Post-classical Cult of Malophoros at Selinus," *AJA* 71 (1967): 335–52. Finally, note the fact that the large tree is called in Hebrew ʾelah, 'goddess' originally. For Egyptian tradition, see Marie-Louise Buhl, "The Goddesses of the Egyptian Tree Cult," *JNES* 6 (1947): 80–97.

29. Helene J. Kantor, "A 'Syro-Hittite' Treasure in the Oriental Institute Museum," *JNES* 16 (1957): 148–49. In fig. 9 the winged disk is supported by the "hieroglyphic Luwian" characters that spell the name of the Hittite king Tudhalias.

30. Ps. 18:11. Cf. 1 Kings 12:28.

31. Heinrich Zimmer, *Myths and Symbols in Indian Art and Civilization* (New York, 1963), pp. 75–76. Mr. Barry L. Ross kindly called my attention to this reference. J. E. Fontenrose has already argued for Near Eastern origins of the Garuda figure in *Python* (Berkeley, 1959), pp. 202–4, 246. Appropriately enough, Garuda is also the name of the Indonesian airline.

32. On the goddess Kali and Canaanite ʿAnat, see the unpublished paper by Marvin Pope, presented at the Biblical Colloquium several years ago.

33. See n. 31, *et passim*.

Fig. 5 Relief of Assur-nasir-pal, showing winged sun disk over tree of life. Early ninth century B.C. British Museum No. 124531. Photograph courtesy of British Museum.

Fig. 6 Winged sun disk over fruit-bearing tree of life. Neo–Assyrian seal. British Museum No. 89135. Photograph courtesy of British Museum.

Fig. 7 Ivory inlay of Syrian origin found in Assyrian palace at Nimrud. British Museum No. 118104. Photograph courtesy of British Museum.

Fig. 8 Two bull-men support-
ing winged sun disk, with human
figure between them. From Tell
Halaf. Ninth century B.C. Aleppo
National Museum No. 16.
Photograph courtesy of Aleppo
Museum.

Fig. 9 Rock cut relief at
Yazılıkaya, near the Hittite capi-
tol of Hattusas at Boghazkale,
Turkey. Probably a goddess
holding the king whose name,
Tudḫalias, supports a winged
sun disk. Probably thirteenth
century B.C.

attack, and the title "lord of heaven," which is also the title given to Horus as the great winged disk in the myth of Edfu.

At the opposite end of the ancient civilized world, we find the tradition of the great winged disk radically modified, not to say misunderstood, in Homer's description of the aegis of Zeus. Professor Hopkins has already demonstrated the dependence of the Gorgon's head upon the solar motifs of the Near East, and he mentioned the winged disk in this connection.[34] The aegis as a goatskin is a mythical readaptation of what were originally bird's wings that "flutter."[35] It was also still associated with the thunderstorm, for

34. Clark Hopkins, "The Canopy of Heaven and the Aegis of Zeus," *Bucknell Review*, December 1964, pp. 1–16. Although this article was of immense value to me, Professor Hopkins cannot be held responsible for the conclusions reached here. See, however, M. J. Mellink, "Excavations at Karataş-Semayük and Elmali, Lycia, 1969," *AJA* 74 (1970): 251–53 and plate 61, for a sixth century B.C. Gorgon from Lycia.

35. Homer *Iliad* XVIII. 204. The association with a goatskin is certainly popular etymology from *aix*, enhanced by the fact that the cult statue of Athena was evidently clad in a goatskin dyed red (Herodotus IV. 189) that was identified with the aegis. Any orientalist will immediately recognize the "goatskin dyed red," but what it meant in the Orient is also completely obscure. To find it as an element of living costume is remarkable, and indicates the survival of extremely archaic custom—unless the royal purple of later times is a continuation. That there is more than one Anatolian cultic element involved in the Homeric aegis is demonstrated by the curious rituals involving the ᵈ*Lama kuršaš* discussed by H. Otten ("Ritual bei Erneuerung von Kultsymbolen hethitscher Schutzgottheiten," in *Festschrift Johannes Friedrich* [Heidelberg, 1959], pp. 351–59), E. Laroche ("Récherches sur les noms des dieux hittites," *Revue hittite et asianique* 46 [1946–47]: 75), and H. T. Bossert (*Ein hethitisches Königssiegel*, Istanbuler Forschungen, no. 18 [Berlin, 1944], pp. 57–58). Since *kuršaš* means fleece as well as (probably) aegis, and is identified with a shield that is regarded as a god to which cult is dedicated, the Medusa (= 'governoress') on the fleece skin or shield of Homeric tradition can hardly be mere coincidence. Further, the *É kuršaš*, i.e., 'house of shields' may very well have some connection with the abode of El in Ras Shamra (*qrš*), and therefore with the tabernacle the covering of which is ramskin dyed red and *tḥš*-skin. Similarly, Solomon's 'house of the forest of Lebanon' was decorated with golden shields of two different types, as indicated by their different names and gold contents, just as the Hittites and Myceneans both had shields of two different shapes, the figure-8 and the oblong (Bossert, *Königssiegel*). Note also the cult of shields, by anointing, as indicated in 2 Sam. 1:21 and Isa. 21:5, and the identification of Yahweh as the "shield" of Abram, Gen. 15:1.

Finally, the curious technical term *qrš* in a very difficult passage of Ezekiel having to do with shipbuilding and parts of ships (27:6) seems to imply a structural feature of a ship made from cypress from the "isles of the Kittiyim." It is more than conceivable that the *qrš* of a ship is specifically what is still known as the "ship's skin," i.e., the outer enclosure of the hull. All these uses and meanings may well derive from a very archaic word for a *baityl*—an object that was ritually identified as a 'house of god,' and that was associated with red-dyed goat- or sheepskin with amazing frequency in ancient cultures. The red garment must certainly be the cultic representation of the "garment of flame"—cf. below.

when Zeus shakes the aegis the storm comes.[36] It is the symbol of his sovereignty that he confers upon Athena, who in turn places it upon the shoulders of Achilles. The ensuing description is most strikingly Near Eastern:

> But Achilleus, the beloved of Zeus, rose up, and Athene swept about his powerful shoulders the fluttering aegis; and she, the divine among goddesses, about his head circled a golden cloud, and kindled from it a flame far-shining. As when a flare goes up into the high air from a city from an island far away, with enemies fighting about it who all day long are in the hateful division of Ares fighting from their own city, but as the sun goes down signal fires blaze out one after another, so that the glare goes pulsing high for men of the neighbouring islands to see it, in case they might come over in ships to beat off the enemy; so from the head of Achilleus the blaze shot into the bright air. He went from the wall and stood by the ditch, nor mixed with the other Achaians, since he followed the close command of his mother. There he stood, and shouted, and from her place Pallas Athene gave cry, and drove an endless terror upon the Trojans. As loud as comes the voice that is screamed out by a trumpet by murderous attackers who beleaguer a city, so then high and clear went up the voice of Aiakides. But the Trojans, when they heard the brazen voice of Aiakides, the heart was shaken in all, and the very floating-maned horses turned their chariots about, since their hearts saw the coming afflictions. The charioteers were dumbfounded as they saw the unwearied dangerous fire that played above the head of great-hearted Peleion blazing, and kindled by the goddess grey-eyed Athene. Three times across the ditch brilliant Achilleus gave his great cry, and three times the Trojans and their renowned companions were routed.[37]

We have, fortunately, a contemporary representation from Assyrian royal art which vividly portrays in visual form what all our sources describe in language. An enameled tile from the reign of the early ninth-century King Tukulti-Ninurta II shows Assur riding in the winged disk with his drawn bow aimed at the enemies of the king (fig. 10). On either side are the storm clouds with rain falling.[38] At a slightly earlier period, a relief depicts the disk with its cloud from which two hands extend, one in a gesture of blessing, and the other handing down a bow which is not drawn.[39] The formal association

36. Homer *Iliad* XVII. 593.

37. *Ibid.*, XVIII. 203–29. That the aegis is intimately related to a thunderstorm theophany in origin is guaranteed by the fact that the word means 'storm,' 'hurricane,' or the like in later times as well. What could be more appropriate to the 'cloud-gatherer,' the wielder of the thunderbolt? See Aeschylus *Choëphoroi* 592. The translation quoted is that of Richard Lattimore, *The Iliad of Homer* (Chicago, 1951).

38. W. Andrae, *Coloured Ceramics from Ashur* (London, 1925), plate 8. The striking similarity in form of the clouds and the female breast can hardly be accidental and may help explain the pre-Mosaic divine name El-shaddai.

39. The "broken obelisk" probably of Assur-Bel-Kala, mid-eleventh century. See K. Jaritz, "The Problem of the 'Broken Obelisk'," *JSS* 4 (1959): 204ff. A very cursory survey of the existing evidence suggests that both the winged sun disk and the heavy emphasis upon the *melammū* enter the Assyrian tradition with the great raid upon the west by Tukulti-Ninurta II (*ARAB*, pp. 128–32), but this needs further detailed investigation.

Fig. 10 The god Assur in a fiery winged sun disk surrounded by heavy rain clouds flies above the king in his chariot. The bottom half is unfortunately broken and lost. From the reign of Tukulti-Ninurta II or Assur-Bel-Kala, and thus either early ninth or mid-tenth century B.C. From Assur. British Museum No. 115706. Illustration after W. Andrae, *Coloured Ceramics from Ashur* (London, 1925), plate 8. By permission.

with later Christian art is striking, but these Assyrian representations also aid our understanding of the bow symbolism in Genesis 9, which has long been a problem since the bow is normally the symbol of war, not of peace and well-being.[40]

The key to the problem lies in the fact that the bow appears in two contrasting contexts in the Assyrian reliefs. The first represents the king attacking the "seed of Tiamat," the unsubjugated enemy beyond the boundary line, and *therefore* the incarnation of absolute evil and chaos (figs. 11, 12). The bow of the king, and also that of Assur's *melammū* above, is drawn and firing at the enemy. It is a most graphic description of the ancient Assyrian "*Blitzkrieg*," complete with storm troops and the raining down of fire (see below), that must end in the "perpetual victory over all enemies."

40. E. A. Speiser, *Genesis*, in *Anchor Bible*, p. 59, where he was unable to correlate the bow with ancient symbolism. Genesis 9 still represents, of course, a radical reinterpretation of very old motifs, for the bow there is not a symbol of the divine power handed down to a political power structure.

Fig. 11 Relief of Assur-nasir-pal II showing the king attacking with the *melammū* of Assur flying above and attacking. British Museum No. 124540.

Fig. 12 Detail of relief of Assur-nasir-pal II showing similar attack of god and king. British Museum No. 124555.

Fig. 13 Relief of Assur-nasir-pal II showing triumphant entry of the god and king. The bow of the god in the winged sun disk is in the stance of the rainbow of Genesis 9. British Museum No. 124551.

The second context is that of the triumphant return. The king parades proudly, bow in hand but undrawn, followed often by the parade of captives and loot which were the real motivation for the war in the first place. Above him is again the king's divine air power, holding his undrawn bow—a mirror image of the all-powerful conqueror. The art expresses even more graphically than the language the fact that there *can* be no conceivable difference of opinion between the policies and operations of the divine state and its ultimate supernatural authority—its "god," which is nothing but the symbol of the state itself: Assur the god *is* Assur the state. (See fig. 13.)

The Flood Story of Genesis 6–9 is a parable based upon this age-old heathen "theology." Granting the terrible evil to which human society can debase itself, yet even (or should we say *only?*) God is capable of regretting the "cosmic storm" which He brought upon the earth, nearly wiping out all life in the process. The undrawn bow of the Assyrian triumph becomes a symbol, then, not of the all-powerful king's glorious victory which enhances the well-being (i.e., the "peace") of all his subjects, but of the determination that never again can the evil and chaos of mankind provoke God into returning the world to chaos (i.e., the Flood) as a just punishment. The undrawn bow is the "sign" of the unilateral covenant by which God binds Himself to guarantee the security even of His enemies from the violence of such overwhelming power that it reduces to insignificance the petty triumphs and proud gloating of "man, proud man, drest in a little brief authority, most ignorant of what he's most assured, [whose] glassy essence, like an angry ape, plays such fantastic tricks before high heaven as make the angels weep" (Shakespeare, *Measure for Measure* II, ii, 116–21).

A complex of traditions and representations associated with the great winged disk have been found from Homeric Greece to India, and from Old Kingdom Egypt to the Ptolemaic temple at Edfu.[41] Were it not for the mythical and epic traditions which have been preserved, it would be rather futile to attempt any further conclusion concerning the meaning of this art motif that has had such far-reaching dissemination in time and space. However, an equally varied documentation in language does exist, and this makes it possible to go much further by correlating artistic and literary sources.

To return to the Assyrian sources of the ninth century, when the winged disk became extraordinarily common, we find frequent statements to the effect that "The awe-inspiring splendour of Assur overwhelmed the men of . . . and they brought tribute."[42] It is intriguing that in the early annals at least the splendor of Assur which motivates the capitulation of enemy kings is almost always referred to in historical contexts that are independent of the king's attack. The attack by the king himself is characteristically described as follows: "For two days I thundered against them like Adad . . . and I rained down flame upon them . . . My warriors flew against them like Zu [the storm bird]."[43]

Fortunately, we have an excellent study by Oppenheim[44] of the words

41. Possibly a related problem is the occurrence of the winged disk on funerary stelas. Beginning in Egypt of the New Kingdom, stelas (fig. 14) of nonroyal personages bear the symbol, and it appears also on a funerary monument in northern Syria about the ninth century. It continues into the Hellenistic period in Phoenicia at least (fig. 15), but in the meantime the winged disk in the form of the Medusa head (fig. 16) became a virtually universal symbol on Greco-Roman sarcophagi (fig. 17).

42. Since Tiglath-Pileser I (*ARAB*, p. 79), the phrase appears (twelfth century); cf. also *ARAB*, p. 113 (Adad-Nirari II): ". . . the effulgence of his surpassing glory consumed all of them. The lands of the kings were distressed. The mountains trembled" (Late tenth century). However, this seems clearly to be a reference to the *melammû* of the king, not of the god Assur.

43. *ARAB*, p. 156. (*Assur-naṣir-pal.*)

44. A. Leo Oppenheim, "Akkadian *pul(u)ḫ(t)u* and *melammu*," *JAOS* 63 (1943): 31–34. See also the monograph of E. Cassin, *La Splendeur divine* (Paris, 1968).

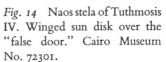

Fig. 14 Naos stela of Tuthmosis IV. Winged sun disk over the "false door." Cairo Museum No. 72301.

Fig. 15 Funerary stela of the Hellenistic period. Beirut National Museum.

Fig. 16 Medusa (Gorgon) head from the Greek temple at Didyma on the southwest coast of Anatolia. As is usual in Greek renderings of the winged sun disk, the two serpents have become a ribbon tied in a square knot under the chin, and the wings have become reduced to appendages emerging from the top of the head.

Fig. 17 Typical gorgon from the "sarcophagus of the Satrap" from Sidon on the Phoenician coast. Dated to the fifth century B.C. The serpents below the wings at the top of the head and forming an overlap knot under the chin are still clearly indicated as such. Except for Egypt and the Phoenician coast, the winged sun disk virtually disappears from ancient art during the early Persian Empire—and at the same time the Medusa/Gorgon becomes virtually universal as a symbol of funerary monuments. It is merely one of a vast number of ancient near eastern cultural traits of great antiquity that rapidly died out or were radically transformed during the Persian and Hellenistic periods. Eski Šark Museum of Istanbul No. 367.

Fig. 18 Detail of headless statue of a Roman emperor, illustrating the *aegis* with Gorgon head and fleece as a symbol of authority. Corinth Museum No. 23385.

Fig. 19 Head of Christ in winged sun disk over the gateway to the Byzantine monastery at Alahan in southern Turkey in the mountains above Silifke. It illustrates beautifully the local preservation of very old motifs readapted to new religious ideology. Readaptation is indicated by the fact that so far as we know the winged sun disk had been moribund for nearly a thousand years in the region. How and why such motifs should suddenly reappear is a question that we can hardly answer. But the process is very familiar to any specialist in the ancient Near East, and we have numerous examples in pre-Greek history of such sudden reemergence of long "dead" cultural motifs.

Fig. 20 Detail of the "Black Obelisk" of Shalmaneser III, showing king Jehu of Israel, or as he is called here "son of Omri," submitting to the Assyrian king. Mid-ninth century B.C. The *melammū* of the god Assur is over the vassal. It is a parade example of the historical functioning of the *melammū* of the god, for Jehu submitted upon his own initiative to the power and prestige of Assyria according to the biblical account—no doubt for purely political reasons. British Museum No. 118885. Photograph courtesy of British Museum.

for 'awe-inspiring splendour' in Babylonian. He has described the meaning of the Akkadian *melammū* and *puluḫtu* as follows:

1. *Melammū* is, first, a dazzling aureole or nimbus surrounding that which is divine.
2. It is shared by everything endowed with divine power or sanctified by a divine presence, for example, weapons, symbols, temples.
3. It is an endowment of the king, the representative and likeness of the gods, which constitutes the divine legitimation of his royalty. It is bestowed on him when he becomes king, but it is revocable.
4. It is also identified with some concrete object worn by the images of the gods, especially in cultic contexts.
5. As a mask, it hides the natural body of gods, kings, and demons. In

medical texts, *melammū* means simply the external appearance of a patient's face.

The twin word *puluḫtu* is similarly:

1. That which inspires fear or awe.
2. A garment of flame surrounding the god or king, probably identified as a red garment or cloak in cultic contexts.
3. Certain uses indicate that the word *puluḫtu* is a linguistic substitute for the person himself.

Since the Assyrians of the Iron Age were using concepts that had a continuous history reaching back to prehistoric times, we find in their art and annals a paradigmatic example of what I shall term "religious identification." It seems to be almost a universal in the history of religion that certain aspects of reality experienced by persons are identified as manifestations of that which is proclaimed in the sacred tradition. Such identification bridges the transcendent and the immanent, the past and the present, the sacred and the secular, the metaphysical and the empirical.

Thus the Assyrian king in certain aspects of his role is identified with the deity Ningirsu; his warriors are like Zu, his *melammū* is the *melammū* of Assur, but the identification is not absolute. The *melammū* of Assur is functional beyond the actual actions of the king. The analogy to the king's two bodies in much later European political theology is about as good a one as I have been able to find. The fact that the *melammū* of the king can be taken from him is clear indication of the distinction between the natural body and the divine nature of the king. Similarly, in the Sumerian king list, the very kingship itself is lowered from heaven and taken up into heaven during the Flood.[45] In the Middle Ages, as early as 875, an interregnum was described as a time when "Christ rules, expecting a king . . . ."[46]

This *melammū* is the word label for a most complex conceptualization of divine and royal glory that is realized in the art motif of the winged sun disk. The *melammū* itself as glory cannot really be seen; all that can be seen is the visible mask with which the *melammū* is identified because the mask has the functional effects necessary to the *melammū*. Since the *melammū* of the gods is associated with a vast range of empirical phenomena—the heavenly bodies, storms, the exercise of royal dominion, war, disease, and so on—it is clear that what later was called *epiphany*, manifestation of deity, was already very complex during the Bronze Age. This "epiphanological polymorphism,"

---

45. T. Jacobsen, *The Sumerian King List* (Chicago, 1939), p. 77, ll. 39–42.
46. Kantorowicz, *King's Two Bodies*, p. 335.

which becomes a very conscious literary motif in Greek mythology,[47] is well illustrated in the Edfu myth: "The majesty of Rē-Ḥarakhte said, 'As thou desirest, O Horus of Beḥdet, thou son of Rē, exalted one who camest forth from me . . .' Horus of Beḥdet flew up to heaven as the Great Winged Disk and therefore he is called, 'great god, lord of heaven' to this day."[48] In another myth of the same series at Edfu, Horus is identified with the acts of five different living beings, at the end of which it is said: "Lo, thou art a flame inspiring fear . . . which lives on in a mound of *kk*-bushes."[49]

Horus is the totality of sovereignty, manifest in the bird which seizes the fish, the lizard which seizes its prey, the hunter's hound which eats the flesh of its catch, the young man who slays the one mightier than himself, and the lion which stands astride the carcass. It is appropriate that he is therefore identified with the winged disk, which is at the same time the symbol of the pharaoh, the incarnation of Horus.

All of our artistic and literary references, therefore, illustrate beautifully and with striking consistency that which can only be described as the deification of power which is recognized in and identified with many forms, but with peculiar clarity in the kingship. It is not surprising, then, that the motif of conflicts between power systems is the most characteristic theme of ancient myth which is also diffused from Greece to India in the Iron Age. We shall focus attention on only one which is of particular interest for our present purposes.

In the myth of the conflict between Baal and Yamm in the Ugaritic texts there has been a long-standing obscurity at the most crucial point in the drama. Yamm sends messengers to the council of gods saying,

> Deliver, O gods, the one whom you obey,
> whom the multitudes obey;
> Deliver Baᶜal and his ᶜanan,
> The son of Dagon, I shall succeed to his *paz*.[50]

No translation of ᶜanan and the parallel *paz* has hitherto commanded general assent, though Driver's suggestion of "lackeys" or related concepts has often been used.[51] The solution now seems to be obvious, but it has been rejected

---

47. Compare the myth of Proteus as only one illustration of the elaboration upon the concept of polymorphism. Note also the various manifestations of Zeus—cloud, swan, etc.

48. Fairman, "Myth of Horus," p. 28 (6, 2).

49. Blackman and Fairman, "Myth of Horus at Edfu-II," *JEA* 29 (1943): 16. The similarity to the burning bush narrative is unmistakable, but inexplicable, historically at least (Exod. 3:2ff.).

50. III AB B, 18–19, 34–35. See H. L. Ginsberg, in *ANET*, p. 130.

51. G. R. Driver, *Canaanite Myths and Legends* (Edinburgh, 1956), pp. 79, 141. J. Aistleitner came very close with the translation *Bote* 'messenger'. In *Wörterbuch der Ugaritischen Sprache*, 2d ed. (Berlin, 1965), s.v.

because of semantic misunderstanding. The thing that can be taken from a god or king that is essential to his status is of course his *melammū*, of which ᶜ*anan* is the semantic equivalent. Its parallel, *paz̄*, is probably to be equated with the Hebrew *kitōnet passīm*, which describes a garment, the traditional "coat of many colors" (like *puluḫtu* in Akkadian) associated with the highest social or political status.[52] In other words, the term does not describe the form of the garment but its social function. It actually is used only of Joseph, who dreams of his brothers bowing down to him, and of Tamar, the princess of the royal house.[53] The fact that both words appear in Ugaritic only in mythical contexts is sufficient guarantee of their extremely archaic vintage, certainly no later than the Middle Bronze Age. To my knowledge, neither word survives in later Iron Age inscriptions.[54]

Unfortunately, all other contexts in which this word occurs in Ugaritic are either broken or completely obscure. Gordon comments merely, "generally appears in contexts concerning deities,"[55] which is accurate as far as it goes. A complete list of occurrences thus far known includes only five other passages.

1. II AB iv 59–60: *pᶜdb.an.ᶜnn.aṯrt*
  pᶜdb.ank.aḫd ulṯ

Whatever this passage may mean, the poetic structure is the well-known ABC:ABD form, which guarantees that ᶜ*nn.aṯrt* and *aḫd ulṯ* are not parallels. The first can only be a vocative: "O ᶜ*anan* of Athiratu." In this case, the ᶜ*anan* is a linguistic representative of the person.

2. II AB viii 14–17: . . . *wnǵr*
  ᶜ*nn.ilm.al*
  tqrb.lbn.ilm
  mt . . .

Here also ᶜ*nn.ilm* is a poetic parallel to and linguistic surrogate for the proper name of the god, Hadad.

"Beware O ᶜ*nn* of the gods;
do not approach the son of the gods, Mot."

52. Contrast Speiser, *Genesis*, pp. 289–90. The contrast between Ugar. *z̄* and Hebr. *s* is simply a shift from a voiced to voiceless sibilant.

53. Gen. 37:3, 23, 32; 2 Sam. 13:18. It is, of course, conceivable that by the tenth century B.C. the word had undergone a semantic shift from a functional designation to a formal identification.

54. But for the survival of ᶜ*ānān* in classical Arabic, see below. Furthermore, for evidence of its survival in popular religion, see chap. VIII.

55. *UT*, p. 458, s.v.

3. Anat iv 76–77: *lk.ᶜnn–ilm*
   atm.bštm.wan.šnt

"Yours is the ᶜanan of the gods; you are powerful, but I am [?]."

It is particularly appropriate that ᶜnn is here correlated with Akkadian *baštu* 'dignity' (*CAD*, s.v.).[56]

4. IV AB ii 32–33:]*h.bᶜlm.dipi.*[
   ]*hd dᶜnn n*[

Although this passage is only a fragment, it seems clear that the ᶜanan of Baal is parallel to the obscure *ipi*, possibly Akkadian *ēpū* 'cloud.' It is particularly unfortunate that the passage is broken, for it seems to be making some statement about the ᶜanan/ipi of Baal/Haddu which might have furnished substantive information about its function or characteristics.

5. 32:4]*at.brt.lb kᶜnn.*[
   "You are pure of heart like the ᶜanan of [. . . .]"

It is especially significant that the root *BRR* which is here an attribute of the ᶜanan is regularly construed with *kšpš* 'like the sun' in the legal texts,[57] and therefore would support further the equivalence of the ᶜanan with the Akkadian *melammū* and the sun disk.

Though very little specific information is yet to be gleaned from these other usages in the Ugaritic texts, every occurrence indicates that the word designates something closely identified with the divine beings: it is a substitute for their names or an aspect of their person. It has not been demonstrated that the word is a semantic equivalent for the Akkadian *melammū*; however, such a working hypothesis goes much further in clarifying the texts we have than any alternative so far suggested. Further, it gives the Yamm epic a vividness and a context in ancient patterns of thought that makes it very attractive. Just as a king's *melammū* can be taken from him, so also in ancient myth a god can lose his "glory" to superior force.

THE MASK OF YAHWEH

It is only when we introduce the very considerable body of evidence from the Bible into consideration that the case for our hypothesis becomes virtually

56. An alternative rendering and interpretation is possible (*UT*). He takes *bšt* to mean 'swift' and *šnt* therefore to be the antonym from an unknown source. In this case the ᶜanan is a messenger emanation for which see below.
57. E.g., *PRU* II 5, ll. 2–4. See Akkadian, e.g., *ki-ma Šapaš za-ki-ti* (*PRU*, III 15.120, l. 14; cf. Cant. 6:10). Is this the Aramean name of the sun god?

a certainty. In turn, the observations concerning the *melammū/ᶜanan* and its parallels from Greece to India throw a new light upon biblical narratives which enables us to place them into a historical and ideological context.

As opposed to the mere six valuable occurrences in Ugaritic, there are eighty-six occurrences in biblical Hebrew, plus one occurrence as a denominative verb (in all probability),[58] and nine uses as a participle (though the participial use may derive from another root). All of the references to the ᶜānān of Yahweh as an active agent in the past experiences of human beings have to do with the Exodus and Wanderings traditions regardless of the literary "source"—L, J, E, D, or P—and the ᶜānān occurs in contexts assigned to all of them. There is no usage of the word in a narrative context after the death of Moses, and the only passages in which it is referred to as any kind of presently empirical reality in later contexts have to do with the Jerusalem cultus.[59] The usages of the word outside the Pentateuch have no characteristics which could possibly be the basis for the peculiar way in which the word is used in the Exodus and Wanderings traditions, though those traditions are frequently drawn upon in the Psalms for poetic restatements. As we have seen, the word occurs in Ugaritic myths in contexts which on linguistic grounds alone must be assigned to a period earlier than the Late Bronze Age. Also, the biblical usages strongly suggest that the word was incorporated into the tradition in the time of Moses, but never again used in that peculiar fashion, though later uses indicate some kind of surviving tradition concerning the ᶜānān. Canaanite-Phoenician traditions evidently did not preserve it at all, to judge from surviving inscriptions (though this conclusion is very hazardous in view of the paucity of our documentation).

It has already been pointed out by other scholars that the ᶜanan of Yahweh is identified with Yahweh himself in the Exodus-Wanderings traditions of Exodus and Numbers.[60] It needs now to be emphasized that

---

58. In the flood story, Gen. 9:14. The translation, "When I bring clouds over the earth . . . ," is completely inadequate, since it does not convey the sense that the ᶜānān is first and foremost a theophany in the form of a storm-cloud. Of course, whether that was still known at the time the present language was fixed is another matter. In light of later uses, it is quite possible that the writer no longer was aware of the original connotations or denotation of the word. See below.

59. I Kings 8:10–11. It survives as a future prediction of doom in the prophets, e.g., Jer. 4:13; Isa. 4:5; Zeph. 1:15; Joel 2:2. In all of these and others, the ᶜānān is still associated with the punitive theophany of God.

60. For the angel manifestations of the Abraham stories, see G. von Rad, *Das erste Buch Mose* (ATD, 1956), "Der Engel des Herrn ist dann also eine Erscheinungsform Jahwes," p. 163; see also his *Old Testament Theology* (New York, 1970), 1:287. This is not true in the present text of the Bible. Historically it is not true, for the narratives of divine epiphanies to man antedating the time of Moses could not have been identified with manifestations of Yahweh until the time of the Monarchy, when the pre-Mosaic

these narratives associate the ʿānān with the exercise of sovereignty. The ʿānān occurs first in the form of an attack upon enemy armed forces as in Exodus 14:24: "And during the morning watch, Yahweh looked down upon the camp of the Egyptians from the column of fire and cloud, and confounded the camp of the Egyptians." It should be noted here that there is but one column, ʿammūd, consisting of both fire and ʿānān; I believe this to be original, the systematization into two columns, cloud by day and fire by night being a later rationalization of a tradition that was probably no longer understood in its original context.[61] It certainly cannot be coincidental that, in the Egyptian Edfu text, Horus "stormed against" the enemy in the form of the winged disk; in the Tukulti-Ninurta enameled tile, Assur, in the winged disk, attacked the enemy below; and in the *Iliad*, Achilles, wearing the aegis, shouted like the trumpets of an attacking army. In all four traditions the result is the confounding and overwhelming of the enemy, in the words of Homer, "and drove an endless terror upon the Trojans."[62] Further, in verses 19–20 of Exodus 14 the ʿānān moved between the fugitives and the Egyptian host as a protection against attack; similarly, the aegis of Zeus warded off the darts of Hector (even though it was dented in the process).[63] As late as Lamentations 3:44, the ʿānān is a covering preventing the penetration of prayers. This passage is very typical of later uses which seem to regard the ʿānān merely as a cloud or a covering, but with the power of manifesting a divine personality almost entirely lost.[64]

---

traditions were superimposed upon the authentic traditions of Israel. What is historical and pre-Mosaic is a priori ancient Near Eastern paganism, granting that that paganism exhibited a vast range of religious phenomena (M. Smith, "Old Testament Studies," p. 27). Cf. also A. J. Wensinck, cited in *CS*, p. 211, and the discussion by Frankfort, who had already identified the winged disk with the "glory" of Assur, in *CS*.

61. I.e., the *melammū* = mask, and *puluḫtu* = fire. It may be coincidence, but Anatolian winged disks often exhibit two columns instead of one as a support. Furthermore, the two bull-men flanking the column, or even supporting the winged disk must be closely associated in the ideology of ancient iconography. In Hurrian culture, one is identified as Šerriš = Dawn (Day) and the other as Ḥurriš = Sunset (Night); this may well be the historical source of the pillar of cloud and the pillar of fire associated with day and night, respectively, in biblical tradition of relatively late date. For the figures themselves, see *Tell Halaf*, plate IX. Compare also the fantastic, almost gnostic, elaboration of unbaptized creatures in the Megiddo Plaque of Hittite origin (H. T. Bossert, *Altsyrien* [Tübingen, 1951], fig. 1115).

62. *Iliad* XVIII. 218.

63. *Ibid.*, XXI. 400ff. Cf. Hopkins, "Aegis of Zeus," p. 3. Compare the Hittite shield goddess, above n. 35, and the literature there cited.

64. Cf. Ps. 105:39 (dependent upon Exod. 14:19); Ezek. 32:7; 38:9, 16. Cf. also J. Levy, *Neuhebräisches und Chaldäisches Worterbuch über die Talmudim und Midraschim* (Leipzig, 1876–89), s.v., for later postbiblical usages.

This very fact, however, combined with many other uses, demonstrates the extremely close analogy to the Akkadian *melammū* and *puluḫtu*, for the *ᶜanan* is the mask, and the fire is the garment of flame called *puluḫtu* in Akkadian. We are of course reminded both of the burning bush tradition of chapter 3 of Exodus, and the Edfu identification of Horus with the flame that burns on in a mound of *kk* bushes and inspires fear.

The great variety which exists in the biblical narratives concerning the manifestations of deity may be introduced by the curious passage of Exodus 14:19: "Then the messenger of the god which was going in front of the camp of Israel moved and went behind them; and the column of the *ᶜānān* moved from their front and stood behind them." It seems quite probable that the two statements are variants of the same tradition, and therefore that *malʾāk* 'messenger' and *ᶜānān* are two different word labels for the same concept: both refer to the manifestations by which deity becomes functional in human experience. A good example is the story of the burning bush; a *malʾāk* appeared to Moses, but it is Yahweh or God who speaks. Yet, it is equally clear that distinction is also made between deity and manifestation, as in the interesting narrative of Exodus 33:2ff. First, Yahweh tells Moses that he will send a *malʾāk* before the Israelites into the land of Canaan to bring them into the land, but Yahweh himself will not go up among them lest he consume them on the way.

Later, in chapter 33, Moses makes a plea that the *pānîm* of God go with them (14–15), which is analogous to the advice of Hushai to Absalom (2 Samuel 17:11) that Absalom himself (*pānêkā*) lead the army against David. But what is the *pānîm* 'face' of Yahweh? The intervening passage of 33:5–11, from the E (Northern) source according to the literary critics, probably is intended to be the answer. "Whenever Moses went into the tent, the column of the *ᶜānān* would come down, and stand at the opening of the tent, and speak with Moses." The people's response in prostration indicates the identification with Yahweh, and the narrative concludes with the statement that "Yahweh spoke to Moses face to face, just as a man speaks to his companion" (verse 11). The story toward the end of the chapter about Moses' request to see the "glory" of Yahweh completes the narrative and at the same time explains why the *ᶜānān* mask must be the form of Yahweh's presence, since no one can see the *kāvōd* (i.e., "glory") and live.[65]

It is quite clear that we have here an enormously complicated development of thought concerning divine theophany, but the conclusions are most thoroughly in keeping with what we know or can infer to be true of ancient

65. The history of this motif is still obscure; the old narrative of Judg. 13:22 is difficult to evaluate historically, but, indubitably, there are very early motifs in the narrative.

oriental thought patterns of the time of Moses. The malʾāk is a "sending," an "emanation," by which the power or word of deity becomes functional in human experience. Thus it is a malʾāk which was sent by Yahweh to deliver Israel from Egypt in the JE tradition of Numbers 20:16, though the narratives of the Exodus emphasize rather the column of ʿānān in the event.

There can be then little question that the 'angel' is the linguistic equivalent of the much older ʿānān. If one observes that the ʿānān of Yahweh is never used in narratives concerning theophanies after the death of Moses, but the traditions concerning the malʾāk continue to the New Testament period, becoming more elaborate as time goes on,[66] then one must conclude that the tradition concerning the manifestation of Yahweh in the ʿānān belongs to the earliest period of biblical history, but no longer is used as a means of independent description of theophanies after the time of Moses. This conclusion is strongly reinforced by the fact that the word does not occur in extrabiblical sources in later Canaanite or Aramaic, and survives in Arabic only in association with demons.

The malʾāk theophanies continue from the time of Joshua (5:13-15) onward, but the Manoah story in chapter 13 of Judges is particularly instructive. First, a malʾāk appeared to Manoah's wife which she describes as a "man of God" whose appearance is like that of a malʾāk (verse 6); Manoah prays for the return of the "man of God" and the malʾāk again becomes manifest to his wife. The malʾāk refuses to give his name, which is peliʾ, and when the flame ascended from the altar on which Manoah had sacrificed a kid, "the angel of the Lord ascended in the flame of the altar . . ."[67] The response of Manoah, who "knew then that he was an angel of the Lord," is classic: "We shall certainly die, for we have seen God." To turn to later biblical traditions, the martyr, Stephen, states that "It was he [Moses] who with the congregation in the desert went between the angel who spoke to him on Mount Sinai and our forefathers, and received and communicated to you utterances that still live . . ." (Acts 7:38). Again, in verse 53, Stephen refers to the law given by angels. These statements are simply incompatible with the biblical account except under the assumption, which seems now to be quite correct historically, that the ʿānān and the malʾāk are actually interchangeable though

66. Compare, for example, the "angels" in Job (sixth century) with Jubilees (third century). Curiously, a god, ʾangelos, was worshipped in the early oracle shrine at Didyma in western Anatolia. Note also the very close relationship between the Greek nikē 'angel of victory' and the angels of Christian art.

67. The tradition that the flame on the altar is a god, or a manifestation of a god, is very old and very widespread: "I manifest myself as the sacred fire that carries the offerings of men on earth to the gods in heaven" (H. Zimmer, Myths and Symbols, p. 44). Cf. "The fire which they kindle it is Marduk, who in his youth . . ." (S. A. Pallis, The Babylonian Akîtu Festival, p. 213 and plate V, 1.3).

they have different nuances and eventually came to have quite divergent meanings.

Fortunately, we again have extrabiblical evidence contemporary with the period of the Judges to control the meaning of the malʾāk. Wen-Amun, in his very vivid narrative report, speaks of carrying an image called "Amun-of-the-Road," a "portable" Amun, which is evidently regarded as a projection of Amun-Re (i 38–41).[68] In 55ff, however, the image is called ʾpwty 'messenger', the divine messenger who accompanies the human one. It is worth observing that the image is kept hidden, housed in what seems to be a tent or other temporary structure erected on the seashore, and constitutes a major ground for Wen-Amun's claims upon the various Phoenician city kings. It is equally noteworthy, however, that the "great god" was rather ineffective in promoting Wen-Amun's mission until an event happened which was identified as the act of the god, namely, the prophetic seizure of a king's attendant.

The malʾāk thus designates something in human experience which is identified with the action of deity. It is a manifestation which is tied to time and place and the capabilities of human senses, but its effects can be recognized and felt elsewhere. Thus the seizure of the royal attendant takes place some distance from the image of Amun, which the attendant has presumably never seen. Similarly, the melammu of Assur overcomes the enemies of the Assyrian king quite apart from the presence of the king or his divine emblems. Here, perhaps, are the beginnings of a concept of divine action which is to develop greatly with time, but which seems to be almost absent from modern understanding. This is the concept that the manifestation of divine power is observable in the behavior of individuals. We shall return to this in order to investigate its foundation in the biblical traditions.

Given this background of conceptualization and artistic realization of religious experience in pre-Mosaic times, we may now turn to the event of crucial importance to the whole history of ancient Israel, including the new Israel of the early church. We have already seen the ʿānān of Yahweh as the means by which the exercise of His sovereignty over the Egyptian host was described, and this has already been correlated with the manifold traditions of the winged sun disk which in turn was a universal symbol of supreme sovereignty in the ancient world. Nearly all of the ancient narratives mentioned described the "overwhelming" of the enemy (the enemy being those with whom there is no relationship by covenant).

This ʿānān also appears in the form of a thunderstorm at Mount Sinai, this time overwhelming the Hebrews who had recently escaped from Egypt through the exercise of the power of the same deity. Curiously enough,

___

68. Gardiner, *Egypt of the Pharaohs*, pp. 306–13 (especially p. 308); *ANET*, pp. 25–29.

at Exodus 19:4, the E narrative even mentions the "eagles' wings" on which the Hebrews were borne to Sinai. The Sinai theophany itself is variously described, but the essentials abstracted from the narratives are the following:

1. The ʿāb of the ʿānān: the cloud of the melammū (Exod. 19:9), with which we may compare the clouds represented by the aegis of Zeus cast about Achilles' shoulders, perhaps rendering him invisible from the head down: "Golden cloud about his head." [69]

2. Thunder and lightning: the voice and weapon of Baal but also of Zeus. [70]

3. Smoke ascended like the smoke from a kiln. The only other place where this phrase appears is in the description of the holocaust at Sodom and Gomorrah (Gen. 19:28). In Homer's description, this has become a signal fire from a besieged city not yet captured, and therefore a flame, not a smoke column. [71]

4. A very loud trumpet blast, a motif unexplained and unmotivated in the biblical account (except by P source, which makes it an "all clear" signal that permits the people to ascend the mountain [Exod. 19:13]). In Homer, the trumpet is a simile to describe the voice of Achilles: "As loud as comes the voice that is screamed out by a trumpet by murderous attackers who beleaguer a city. . . ." [72]

5. Moses spoke, and God answered in thunder; Achilles shouted "and from her *place* Pallas Athene gave cry." [73]

6. All the people in the camp trembled (Exod. 19:16). Athene and Achilles, through their shouting, "drove an endless terror upon the Trojans." [74]

We can, in view of the astounding recurrence of motifs,[75] place the Sinai theophany into the framework of our winged disk tradition, which must have been current in literary motifs or epics somewhere in Canaan or

---

69. *Iliad* XVIII. 205–6.

70. *Ibid.*, XXI. 401.

71. *Ibid.*, XVIII. 205–12.

72. *Ibid.*, ll. 219–21.

73. *Ibid.*, ll. 217–18. Note the technical usage of *māqōm* 'place' also in biblical Hebrew. But what does it mean in Homer?

74. *Ibid.*, l. 219.

75. To this can be added a curious similarity concerning the composition of the divine carpetry associated with a theophany. According to the Bible, the floor was "as it were a pavement of sapphire stones, like the very heaven for clearness" (Exod. 24:10 (RSV)). According to Greek tradition, the floor in Zeus's palace was laid with plates of gold (Nilsson, *A History of Greek Religion* [Oxford, 1949], p. 146). Compare Rev. 21:21 where the *streets* of the heavenly Jerusalem in good democratic fashion are pure gold, "transparent as glass."

Egypt during the Late Bronze Age, though I have not yet been able to find it.[76] It seems incontrovertible that the Sinai theophany was described with language which was already current long before the tenth-century reduction to writing, and it is most tempting to conclude also that the present narrative was written (or expurgated perhaps) by someone who did not understand the symbolism. Homer characteristically reduces a traditional description of a theophany to make a rousing good epic narrative.

The thunderstorm is an appropriate base for the description of a theophany because it had been a universal natural sign for thousands of years prior to Moses. Since the column of smoke is also a motif in the description of theophany, it is no longer necessary to search for an active volcano as the proper candidate for Mount Sinai. Nor should the entire narrative be dismissed as mere survival of epic poetry with no historical basis; the Sinai event was precisely the coincidence of known description of theophany with an actual terrifying thunderstorm—a classical illustration of "religious identification" according to my previous definition. It will not, therefore, do at all to dismiss the Sinai event as a mere theophany: we are faced with the historical problem of understanding why this theophany had such a profoundly different and permanent effect upon the whole subsequent history of religion to the present day.

The solution to the historical problem of Sinai must lie in the recovery of a more adequate context within which to examine the event. The generations-old idea that the religious significance consisted merely of a change of god on the part of a primitive nomadic tribe and the establishment of a personal relationship between the god and the tribe has recently been reaffirmed by McCarthy in a careful literary analysis of the Sinai traditions.[77] This is not the time or place for either polemic or detailed refutation; indeed, such a confrontation is not even possible, because there exist no Marquis of Queensbury's rules according to which the debate can proceed. From my point of view, the message of Moses and the emergence of that religious community of the Early Iron Age which was called Israel were inseparable from the

76. But see the description of the fiery manifestation (a meteorite?) in the heavens that appeared to the Hittite army on the occasion of Mursilis's invasion of the Arzawan country and then swooped down on the city of Apasas (= Ephesus?), mortally wounding the king of Arzawa. See A. Goetze, "Die Annalen des Mursilis," *MVAG* 38 (1933): 45–51, and W. J. Pythian-Adams, *PEQ* (1946), p. 119. The Hittite texts say in such cases, "The god showed his *para ḫandandatar*," a term which is equated by the ancient scribes with Babylonian *mêšarum* 'justice.' As Goetze very rightly points out, the equation should not be taken too seriously. In view of the evidence accumulated here, the Hittite concept must be extremely closely related to the later Assyrian *melammū* and the 'lord of heaven' tradition, particularly if the word is related also to Hittite *ḫanda* 'face' (= West Semitic, *panîm*).

77. *Treaty and Covenant*.

predicament of the civilized world of that time. The chronology of events is still uncertain, but it is quite clear that the formation of a new religious community either shortly preceded or immediately followed the complete collapse of political organizations. I much prefer the latter alternative for reasons of political and military logistics.[78] Analogies from the history of religion strongly support the same conclusion. At the present time, archaeological investigations have proven only that both theories are possible, depending upon the working hypothesis accepted.[79]

The analogy from the history of religion would strongly suggest that a religious message formed the basis of a community much broader than any that had previously existed. The immediate and most urgent problem for mankind is that of forming a larger community from the disintegrating fragments of small interest groups, whether familial, tribal, small urban, greater urban, or national. Historically, the size of the interest group is a matter of no particular importance. A primary social function of all world religions which originated through a break with past tradition was to *create* a community, as Montgomery Watt said of early Islam, for those who had no community, or whose community had disappeared from beneath their feet.[80] I should add, for those whose community was no longer adequate to meet the stresses placed upon them by rapid change.

The context of early Israel was one of disintegration of power structures; the mid-thirteenth century saw the whole civilized world divided among four great empires. After less than two generations, little was left of any of them; widespread destruction had taken place from Troy to the borders of Persia, and the toll in human life must have been incalculable. An economic and political dark age set in which was to last for over two centuries; but it was during this period that the new religious community called Israel developed its distinctive patterns of thought that were normative even for early Christianity over a thousand years later.

The Sinai event and the covenant established there are the only reasonable sources for those aspects of Old Testament thought which constitute its amazing contrast to the tired old political theologies of ancient paganism. Our investigation of the winged sun disk leads to the conclusion that it represents symbolically the manifestation of the Divine in the human world —anything that has a divine "glory" or *melammu*. It is not surprising that the symbol is most closely associated with the king who is the supreme manifestation of the divine in the human world, namely, the monopoly of force

78. See Glock, "Warfare in Mari and Early Israel."
79. For an exhaustive survey of recent publications, see *LIS*.
80. Montgomery Watt describes the social process in *Muhammed at Mecca* (Oxford, 1953), pp. 151-53.

which operates through the usual functions of war, law, and economic control. In other words, the old paganism proclaimed that the supremely important factor in human history, the source of all security and economic well-being for the common run of humanity, is the power of the political state which manifests in clearest form the nature and being of the gods.[81]

The extent to which this deification of force drove out the authentic religious tradition of the Old Testament stemming from the time of Moses is illustrated vividly by the fact that shortly before the destruction of the state of Judah, the winged sun disk became the symbol of the king, probably, as Tushingham has recently argued,[82] during the reign of the "righteous" king Josiah. The symbol could have meant nothing other than the fact that the king of Assyria had lost his *melammu*, at least in Palestine, and that the king of Judah had received it. But in the process, Yahweh who conferred the *melammu* was reduced to not much more than the supernatural symbol of legitimacy exactly like the god of Assur. Josiah lost his life because of this heresy, and Judah lost its corporate existence shortly afterward because it could not conceive of the idea that God (or should we say the historical process?) might have other concerns than the preservation of a thoroughly paganized group of ambitious politicians. The prophet Jeremiah nearly lost his life in the confrontation of the two opposing ideologies, and the violence of the reaction to his message is the best indication of the fact that his contemporaries placed very little value upon the authentic religious tradition for which he pled in vain.

What happened at Sinai, then, represents at least a temporary recognition of the fact that political power structures cannot claim to be the "Ultimate Concern," to use Tillich's terminology—that they cannot be any manifestation of God at all. The rejection of the divinity of the pharaoh and of his authority entails the rejection of any political power structure as a religious concern which takes precedence over ethical and moral principle. At Sinai there was made real in human experience the fact that there was an alternative to the deification of the state as the infallible cause of human well-being and security, and the final arbiter of human obligation. It is difficult to imagine a more perverse and vicious ideology than the notion that ethical obligation is a mere function of a social or political boundary line, but, unfortunately, there is little evidence of any other concept in the

81. For further discussion of this motif, see H. Frankfort, *Kingship and the Gods* (Chicago, 1948), pp. 295–312. For a clever restatement of the old pagan ideology in modern terms, see Cox, *Secular City*, esp. p. 254: "... the mode of theology which must replace metaphysical theology is the *political* mode." Metaphysical theology is usually the consequence of politics, not its contrast.

82. "A Royal Israelite Seal(?) and the Royal Jar Handle Stamps," *BASOR* 200 (1970): 71–78; 201 (1971): 23–35.

ancient world. One may well add the modern as well, in view of the ideologies and propaganda of the polychrome "power" movements, all of
which exploit to the point of insanity arbitrary and accidental contrasts
between human beings for the purpose of establishing or maintaining a
power system.

The ancient as well as the modern paganism was not lacking in occasional expressions of noble philosophical ideas. Perhaps these are always
much more attractive than the prosaic humdrum of historical reality, yet
even ancient man could recognize that lofty religious principles are mere
"mighty affirmations of words"[83] until they become illustrated, not by
aesthetic creations of artists, but by deeds. Sinai represents the brute reality
of human experience in a thunderstorm, a theophany following upon a
human experience of liberation, correlated inseparably with the absolute obligation, voluntarily accepted, to translate obligations couched in human language (and therefore abstractions) into deeds, especially to refrain from acts
that are incompatible with any stable social life—the Ten Commandments.
The transcendent value of what may seem to us rather trivial and common
experience is thus inseparably bound up with the equally trivial and common
kind of morality which is nevertheless necessary for the existence of any
tolerable social life: the security of persons from attack, the good faith and
honesty between persons in all kinds of negotiations, the love and respect
between the generations of humanity, the security of family relationships,
and freedom from aggression against the structure of economic and social
functions upon which all civilized man is dependent. Perhaps it is only when
these fundamentals of social life become unpredictable that they can be
properly valued. This is what Sinai meant; the community formed there
accepted these not as God-given rights, but as God-given obligations to
which they in effect pledged their lives as guarantees. Before two centuries
had passed, the newly formed community found ways in the pre-Mosaic
paganism to substitute rituals for obedience under the pretext of political
necessity. The consequence was division in 922, destruction in 917, 722, and
597, and finally the mass suicide of Masada in A.D. 70. Fortunately, the latter
did not represent the sane opinion of most Jews of the time who evidently
felt that the rule of God had nothing to do with the madness of power
politics.

83. Wen-Amun, in *ANET*, p. 28.

EXCURSUS

THE MIRACLE AT ZAZ: CHRISTMAS 1966

Zaz is located 20 miles northeast of Midyat in southeastern Turkey. The following account was summarized from a description of the incident by the priest of the Zaz Syrian Orthodox Church, made in the presence of the villagers and tape recorded by Jack Bazalgette who accompanied the Reverend Melvin Wittler to Mardin province in May, 1967. Jack Blake and Mrs. Bazalgette completed the party.

Wednesday, May 17. Having returned to Midyat for the night, we began traveling early the next morning by jeep with Kerburan as our destination. However, en route we stopped at the village of Zaz which was the site of a reputed miracle in January of this year. In Zaz we met first the Syrian Orthodox priest who took us to the church and described the miracle as follows:

On the night of the third day after Christmas last year a woman named Meryem called people from the village to her home and began to relate to them story after story. Meryem had been bedfast for over two years, during which she had not been particularly communicative with her neighbors. The nature of her illness had not been diagnosed. As she was relating stories to the people gathered around her bedside, she told them that unless the fighting in the village stopped, a great calamity would befall the village. She said that a light was burning in the church, but as the people questioned her, she insisted that they all go to the church to see. After not having been out of bed for two years, Meryem joined her neighbors and walked to the church. As they approached the church there was the appearance of a bright light shining through the walls, there being only a very small window near the arched ceiling of the stone church. When the community entered the church, the candles on the altar were lighted and a silver overhead lamp was aflame, with oil overflowing on the floor below. The whole village was amazed at the sight and began to collect oil from the overflowing oil lamp which continued to flow for a week or ten days. Word of this event quickly spread to all the surrounding towns and villages and people came from great distances to collect oil from the flowing lamp. The priest reported ten thousand bottles of oil had been collected, and one of those bottles of oil was shown us in the home of one of our hosts in Midyat. The morning after this event several men from a village four or five miles away came to inquire what damage had resulted from the fire of the previous night which they had seen from their village. The feuding which had gone on for years stopped and the church is now in the process of being repaired by the villagers themselves, who had before been unable to agree on the necessary repairs.

67

We saw the villagers actually doing the building renovation on the day we visited the church. All were working together most harmoniously. We also visited the one-room home of Meryem, who talked with us although she did not want to discuss the event. She and her husband with their two or three children continue to live together, but they live as celibates which she feels they must since she considers herself a holy woman. At the same time she feels responsible for the children whom she has brought into the world. We were in no position to verify the facts of the event, but we certainly were impressed that a change in the life of that village had been made as a result of something's having occurred.

III

THE "VENGEANCE" OF YAHWEH

It has long been taken as axiomatic that the primitive desert institution of the blood feud was a social characteristic of early Israel.[1] In part, this conclusion is due to the assumption that early Israel and the conquest of Palestine represented the historical process, attested in later periods of the Near East, of gradual sedentarization of desert nomads. The classical treatment of blood vengeance by Merz in 1916 well illustrates the position which has not really been effectively challenged by later generations of scholarship:

> There have been scholars who denied almost indignantly that the custom of blood vengeance ever was characteristic of the Hebrews. . . . Today scholarly work is freed from those tendentious presuppositions. It has recognized that the Israelites did not emerge into history from the beginning with a closed and ideal culture, but rather had to go through a long development which led them from primitive beginnings to their later brilliance. Therefore it is obvious that at some specific stage of the process the institution of blood vengeance was at home, which can be established for all other peoples of the earth.[2]

He went on to observe, quite correctly, that there is not a single classical instance of blood vengeance in the Hebrew Bible.[3] The material is extremely meager, and consists altogether of cases which have some exceptional character, or which happen in the highest social circles, particularly the royal court.

A corollary to this assumption that early Israel was a primitive tribal group with primitive customs has been the idea that Yahweh was also a primitive god characterized by "wrath and vengeance," which contrasts sharply to the more enlightened concept of deity in the later prophets and especially the New Testament. Though many theologians have rightly

---

1. A recent statement of the classical position in de Vaux, *Ancient Israel*, pp. 10–12.
2. Erwin Merz, *Die Blutrache bei den Israeliten*, *BWAT* 20 (1916): 1.
3. *Ibid.*, p. 70.

protested this gross distortion of historical reality, there has never been an adequately grounded study of the word uses and their corresponding patterns of thought which have been thus misunderstood.

It is true that social institutions and customs are closely linked with social organization, and both in turn are linked with religious conceptualizations about the nature and actions of deity. The thesis presented here, however, is that if we analyze the actual word uses that have supported the ideas of blood vengeance held by many modern scholars, the results are simply incompatible both with ideas of primitive tribal organization and the concept of God that have long been considered to be self-evident. Indeed, the only conclusion that can be justified upon the basis of the biblical, and indeed the prebiblical ancient oriental usages of language, coincide with a high degree of precision with those reached in the study of the ᶜānān of Yahweh. If the latter is the means by which the invisible deity becomes manifest to the world of human beings, that which is called the "vengeance" of Yahweh actually designates those events in human experience that were identified as the exercise of the sovereignty—what the ancient Romans called *imperium*.

BLOOD FEUD AND SOCIAL ORDER

The blood feud may best be regarded as a private war, and as such seems throughout the history of man to have been incompatible with any higher kind of social order. It exists where no orderly, legal procedure is available, especially between groups that have no social relationship to regulate their disputes. The blood feud, then, will not be characteristic procedure for internal disputes within a tribe. To be sure, there have existed societies within which the defense of absolute moral rights by self-help was tolerated. Indeed, no society has ever been able completely to put an end to private vengeance, and certain types of private force for vengeance have been at times systematically condoned by society: "What is revenge but courage to call in our honour's debts?" said one eighteenth-century English author.[4] Against this, however, we may cite Cicero as a more characteristic representative of civilized society: "Nihil tam contrarium iuri et legibus, nihil minus civile et humanum, quam, composita et constitua republica, quidquam agi per vim" (*de legibus* 3.18, 42). Elsewhere he states (*pro. T.Annio Milone*, 13) that there can be no action of violence between citizens which is not against the republic. The same position is of course maintained in modern legal theory[5] and is virtually identical to that which Rudolph

---

4. Young, *Revenge* I (1721): i. Cf. *The Oxford English Dictionary*, vol. 8, s.v.

5. When Roscoe Pound speaks of "regulation of self-help and self-redress instead of a general prohibition thereof..." as a feature of primitive law, he may be introducing

Sohm describes as true of Roman law:

No man need submit to being forcibly and without authority deprived of what belongs to him. In repelling any such attack on his property, he is merely protecting his right by his own force.... The person assailed may be said, in a sense, to be exercising the right he is defending. But it is a different matter if the violation of the right is past and complete. It is not then a question of preventing, but of redressing the violation of right which has already taken place. In this case private force, or self-help, is not allowable. To attempt to obtain redress by means of your own strength, would be, not as in the first case, to exercise, but to transgress, the private right which has been infringed, because private law only confers rights of dominion over material (or unfree) objects, and never confers any direct power over the free will of an independent person. To coerce any will which offers resistance to the law, in other words, to execute the law, is in Roman as well as in modern law, reserved for the state. Once a right has been definitely infringed, there is only one way of securing an execution of the law, and that is by invoking the power of the state, in other words, by an action at law.[6]

Though we do not, as usual, have any systematic description from the ancient Orient of attitudes toward actions of self-help or private vengeance, it seems quite probable that Cicero's description would be just as applicable to ancient Egypt or Babylonia or early Israel as it was to ancient Roman legal theory. There seems to be no word in any of the ancient languages that designates the blood feud as an *accepted* social institution, and though actions of self-help certainly existed then as they do now, they are invariably regarded as crimes against society, with an exception to be noted below. The words usually translated as "vengeance, avenge" occur almost exclusively in contexts of mythology and international politics. Thus Horus is the "avenger" (*ned*) of Osiris,[7] and Marduk "avenges" (*gimillam turrum*)[8] in the Creation Epic. In the Assyrian royal inscriptions, the king evidently represents himself in the role of Assur in the Creation Epic, "avenging" the loyal subjects of Assur over against the forces of evil, seed of Tiamat.[9]

---

a useful distinction for some societies, but it is an irrelevant one to the ancient Near East. The historical situation is more complex than a nice legal theory. *An Introduction to the Philosophy of Law*, rev. ed. (New Haven, 1954), p. 33.

6. Rudolph Sohm, *Institutes of Roman Law* (Oxford, 1892), p. 147.

7. Cf. *AH*, s.v. *nd*, where they translate "schützen"; as a title of Horus: *nd itf* "Beistand seines Vaters."

8. *ANET*, p. 66, tablet IV, l. 13, where the "avenger" receives "kingship over the universe entire." Cf. also tablet VI, l. 163.

9. E.g., Sargon "avenges" Tarḫulara, whose own son slew him (*ARAB* II, p. 13). Cf. also p. 5 *et passim*. The 'seed of Tiamat' (*tabnīt Tiamat*) as a designation of the enemy of the Assyrian king occurs in the inscriptions of Assurbanipal (*ibid.*, p. 385), where it is in apposition to 'image of [the *gallū*-demon],' which likewise appears elsewhere. For the identification of "Tiamat" with anything beyond the Assyrian control, compare

In summary, the words for "vengeance" even in ancient pagan texts are used characteristically only of actions carried out by the highest of social and political authority—the gods and the king—and the action is virtually restricted to warfare beyond the body politic. It is true that myth tends to preserve reminiscences of very early or prehistoric social custom, but there is no body of evidence known to me that would justify the conclusion that any state in the ancient world knew or tolerated an institutionalized blood vengeance like that of the Bedouin. If the thesis concerning early Israelite blood vengeance rests upon the assumption that the social structure was essentially that of a Bedouin tribe or group of tribes, then the argument from Assyria or Egypt is untenable. Quite the contrary is true: a study of the various uses of the Hebrew/Canaanite root *NQM* demonstrates an almost precise analogy to the uses of its semantic equivalents in other ancient languages, but with most significant contrasts which well illustrate the peculiar sophistication of ancient biblical thought. Furthermore, those uses illustrate with particular cogency the main thesis of this collection of essays, namely, that early Israel can be conceived of as a functioning social organism only as the actual dominion of Yahweh. As opposed to the currently fashionable dualism of religious conviction and sociopolitical reality,[10] ancient biblical life was characterized by a constant "feedback" between the two areas of human experience. The link between the two was, first of all, the experiences identified as divine action, and, second, the whole complex of religious obligation (in other words, the covenant). It follows, then, that any word that has religious associations is likely a priori to have historical implications for the history of culture, institutions, and law, and the reverse is also true. The rejection of historical controls on theological speculation simply brings about a systematic misunderstanding of both biblical faith and biblical history. On the other hand, the investigation of religious thought in the Bible can and should have important historical connotations. This is not to deny, of course, that the preservation of archaic forms in the cultus particularly is characteristic of most religions. But when a form becomes restricted to the cult, its historical and social function is inevitably minimized, and other social forms become operationally dominant. Such a situation is

---

also the building inscription of Sennacherib (*ibid.*, p. 188, para. 447). Cf. also Sefire Treaty III, ll. 11, 12, where "vengeance" is an act *required* of a vassal if the suzerain is murdered.

10. Note also the very rare application of mythical concepts to historical narrative in the Assyrian royal inscriptions. The art represents Assur "proceeding into the midst of Tiamat," but the historical narratives do not seem to designate the king's campaigns in such terminology. After all, if the campaign is successful, the land concerned is no longer "Tiamat."

not generally true of a formative period in a religious community, but of a period of traditionalism, when it is less a matter of living religious convictions controlling behavior in a free (and even amorphous) society, than social conventions accumulating in an increasingly rigid social structure. The cultus then operates primarily to uphold the social structure, not to furnish qualitative controls over individual behavior within the society, much less the corporate behavior of the structure.

To put it another way, there is little religious content in the substantive law, and religion becomes primarily associated with procedural, adjective law in that its primary function is to uphold the existing legal authorities. This situation seems to have prevailed in the Canaanite society (as well as the Egyptian) in the period before Moses, and it forms the backdrop of religious complexes against which the biblical society was both a contrast and a revolt. In view of that contrast, it is all the more impressive that biblical Hebrew in earliest time used the root *NQM* in almost exactly the same way, and with the same grammatical constructions, as those usages attested in prebiblical West Semitic. Of all attested uses of the root *NQM* from the Execration texts to pre-Islamic Arabic, only two could have any connection with the blood vengeance: chapter 4 of Genesis and a fragment quoted from one of the Mari letters, without context as yet.[11]

The root with which we must deal is among the oldest in the meager inventory of West Semitic words available from the Middle Bronze Age. As early as the Execration texts, the root occurred in one of the proper names, *nqmwp^{c}i*,[12] which can hardly be separated from the Ugaritic royal dynastic name, *Niqmepa*. The fact that the religious associations of the root were already fixed, not to say stereotyped, long before the time of Moses, is virtually proven since the same two roots are still associated in Psalm 94:1, *ʾēl nᵉqāmōt YHWH, ʾēl nᵉqāmōt hôfiᵃᶜ*, where the archaic ABC:ABD form further attests the antiquity of the phrase. The prebiblical association of the root with the actions of the gods is guaranteed by the theophoric names like *Niqmad*:[13] *niqmi-haddu*, and perhaps the Hurrian deity *Niqmis*, which should be regarded as the deification of an abstraction,[14] a characteristic constant of early historic or late prehistoric religion in the ancient world.

---

11. C.-F. Jean, "Les noms propres de personnes," in *Studia Mariana*, ed. A. Parrot (Leiden, 1950), p. 87. *be-el ni-iq-mi-su* (sic!) *i-du-uk-šu*.

12. Georges Posener, *Princes et Pays d'Asie et de Nubie* (Brussels, 1940), p. 65.

13. *PTU*, p. 168. The only names from Ugarit are the two already mentioned, and both are royal dynastic names. On this evidence, the root, even in proper names, seems also to be an archaic survival from inland (i.e., "Amorite") origins. Cf. *APN*, p. 242.

14. For the deification of abstractions, compare, Egyptian *maᶜat*, Akkadian *mêšarum*, Greek *dikē*, and so on.

In view of the one attested occurrence in a Mari document, *bēl niqmišu idūkšu*,[15] and its frequency in Amorite proper names,[16] the conclusion that the root's origin was in the inland West Semitic area, not the Canaanite coastland proper, is justified. So far it has not been attested in Ugaritic other than in proper names. The reference of the root in one of the Mari documents is impossible to determine in the absence of the context, but it would be naive to conclude that it can only refer to the blood feud. The parallelism of *bēl niqmi-* and *bēl dabābi* suggests some technical legal significance. My purpose here, however, is not to argue that the root never had an association with the blood feud—indeed the biblical narrative in chapter 4 of Genesis would indicate the opposite. Rather, my point is that our sources are in all probability self-classifying, in the sense that a society sufficiently developed to have produced these written documents certainly would also have transcended the blood feud as a socially acceptable means of obtaining redress.[17] Consequently, the semantic range of the word underwent a very significant shift which scholars have heretofore overlooked. In other words, where the blood feud still existed, there is no documentary evidence of it until the pre-Islamic Arabic inscriptions of the desert regions.

Though, of course, it is not possible to cite conclusive evidence for the early period, the semantic shift can very plausibly be reconstructed as follows: the blood feud as a private war which entails the taking of an innocent victim's life (often), simply because he belongs to a tribe one of the members of which has committed a murder (corporate liability), always seems to be associated with religious sanctions. This is very clear in Safaitic inscriptions; a god is the ultimate ground of authority for engaging in the feud, as the god is also conceived to be the power which gives success to the venture.[18] In this reconstruction, the story of Cain (chapter 4 of Genesis) and the Safaitic inscriptions reinforce each other very well, but it needs to be

15. Cf. above, n. 11.

16. *APN*, pp. 241–42.

17. Merz makes this same point: "Weil das Volk Israel erst nach seiner Sesshaftwerdung in das volle Licht der Geschichte tritt, so werden wir von vornherein erwarten, die Blutrache nur noch in modifizierter Gestalt anzutreffen" (*Die Blutrache bei den Israeliten*, p. 70). It is thus quite evident that Merz's entire thesis is dependent upon the purely arbitrary assumptions that (1) there was a social organization of Israel prior to its appearance on the historical scene; and (2) that it was a nomadic, barbarian, primitive society. Given this arbitrary nature of both theses, the entire discussion is therefore devoid of historical foundation.

18. *SI*, no. 442, p. 118. Cf. also R. Dussaud and F. Macler, *Mission dans les régions désertiques de la Syrie moyenne* (Paris, 1903), pp. 274, 553, and Mark Lidzbarski, *Ephemeris für semitische Epigraphik* (Giessen, 1902–15), 2:350. The word for blood vengeance is however *t'r* not *nqm*. Early Greek thought was no different—see the story of Hermotinus' revenge upon Panionios (Herodotus VIII. 105–6).

emphasized that from the point of view of the biblical writer, the blood feud was characteristic of the period before the Flood and was escalated into intolerability in the famous Song of Lamech. There is no way in which this narrative can be construed as evidence of blood vengeance in early Israel. On the contrary, it may be a reflection of and reaction to the blood feuds of the desert tribes, especially the Kenites.

As argued above, the establishment of a state which incorporates a number of tribal groups means that the authority and power which formerly sanctioned the blood feud has been delegated by the gods themselves to the state, i.e., the king.[19] Consequently, any appeal to the gods to sanction a privately determined feud within the state is a challenge not only to the royal authority, but to the gods as well. This is merely one of many facets of the various concrete realities of life which virtually demanded an identification of the king or the state with the gods: there could be no state of conflict between the two to which individuals or groups could appeal to justify unilateral action on their own.[20] At the same time, the king either creates or gives sanctions to the courts which settle disputes, bring them to an end, by orderly nonviolent means. In the process of such development, which nearly always *precedes* the production of written documents, i.e., is prehistoric, the divine sanctions associated with the actions called NQM remain, but are primarily if not exclusively prerogatives of the king and state, and it is true that a very large percentage of proper names which include the root are those of royal personages.[21] Instead of representing merely a primitive custom incompatible with any stable peaceful society, the root NQM has to do with the very foundations of political legitimacy and authority long before the time of Moses. Such a dynastic name as Niqmi-Haddu would then be translated into modern terms in some fashion such as this: "The great god Hadad is the ultimate social and historical power which undergirds my authority and legitimate power to act." The old Roman distinction between *officium* and *imperium* is particularly useful in this connection, because it brings out what is most characteristic of Semitic NQM: the *legitimate* power to act.

According to Düll, the Vindex (from which of course we derive both vindicate and vengeance) was a person actually empowered to settle a

19. Logically, then, early Israel as the kingdom of Yahweh must have meant that the blood feud was impossible; the word uses described below prescribe precisely this conclusion.

20. A fortiori, if Yahweh is king, any appeal to a god for a sanction of blood vengeance would have to be to a foreign god—and therefore both polytheism and treason combined. Cf. Exod. 15:18.

21. There is not one personal name in biblical Hebrew that is clearly composed from this root.

controversy between two contenders.[22] Under Roman law the Vindex eventually became an official, the Praetor. In the ancient Near East, he became the king, whose function combined the executive and judicial functions, though probably exercising the latter only as a last court of appeal. It seems very probable that Greek terminology reflects a similar historical and social development: ꜣekdikéō meant originally "to put an end to a controversy," and only at a fairly late time did it degenerate, like vim dico, and nāqam, to mean simply private vengeance.[23]

The various uses of the root ŠPṬ exhibit a very similar sort of semantic development. In the Mari period, ŠPṬ refers not to a judicial act of an officer in a court of law, but to an authoritative administrative edict of a person who exercises an imperium.[24] This use certainly survives in the Phoenician suffete as well as the early Israelite šōfēṭ, and sōfēr—the one originally "sent" to carry out a function with which he was charged by competent political or social authority.[25] Since, as will be shown below, NQM and ŠPṬ are virtually synonymous in the early period, it is curious that the participle form of NQM never occurs as the designation of some sort of social authority. Why this should be is difficult to say in view of the vindex of Roman law: it seems probable that the term was so closely associated with the imperium of God himself that transference to a human authority was consciously or unconsciously resisted.[26] The participle is used of God only in late sources (Nah. 1:2; Ps. 99:8), or of the instrument of God in punitive action (the curses) for breach of covenant (Lev. 26:25). Interestingly, in the latter passage, there may be a reminiscence of the sevenfold vindicatio of Cain.

To summarize, all evidence points to the fact that the root NQM is used in situations calling for the exercise of force in contexts that the normal legal institutions of society cannot handle. It refers to executive rather than judicial action, but it is always either clearly based upon some sense of legiti-

22. R. Düll, "Vom Vindex zum Iudex," *Zeitschrift der Savigny Stiftung für Rechtsgeschichte* 54 (1934): 98–136; 55 (1935): 9–35.

23. It is curious that a local official termed the ꜣekdikēs existed in late Roman times in Syria as a go-between who was responsible for relations between the local community and the central government. Further study is needed in this area. See L. Robert, *Noms indigènes dans l'Asie-Mineure gréco-romaine* (Paris, 1963), 1:472, 487.

24. See Speiser, *Genesis*, p. 134.

25. Judg. 5:14. Cf. also the place name Kiryat-sēfer, which must certainly be read K.-sōtēr, in view of the fact that its other name is Debir, a popular etymology for D/Tapara: 'governor?' See chap. VI.

26. Even more cogent, perhaps, is the fact that the root is disused in Ugaritic and all but absent except for two personal names even in the onomastic tradition. The semantic value of the root emerges into historic significance only at periods of crisis even in biblical Hebrew—and New Testament Greek. See below.

macy or is actually the prerogative of the divine world which of course normally is delegated to the political institution. To revert to the analogy from Roman political theory and practice, an act designated as *NQM* is one which constitutes the actual exercise of an imperium. It cannot, therefore, be equated with vengeance defined as the exercise of private self-help; indeed, it is virtually the opposite, though like private self-help, it does transcend normal legal procedures. The uses of the word correspond almost precisely to the various uses, past and present, of *vindicatio*.[27]

DEFENSIVE VINDICATION

Extrabiblical Sources

According to Düll, the theoretical vindex as avenger had already given way to the vindex as defender before the earliest legal sources.[28] A similar development seems to have taken place during the course of the Bronze Age, but in a different way, for in the Near East the *vindex* as such never really became a sociolegal function,[29] at least not of private citizens. Instead, in pre-Israelite Palestine the defensive vindication is already a function of the highest authority in situations of peril in which there is no other predictable source of help, including self-defense.

The Amarna letters provide the sources for this conclusion and constitute the sole evidence so far of the prebiblical use of the root *NQM* in a

27. Compare the discussion in any unabridged English dictionary, with especial attention to archaic and obsolete usages. They may be classified as: "protect, avenge, punish, claim." All but the last are attested in biblical and prebiblical usages, and the last may be attested in Safaitic inscriptions. See below. It is difficult to imagine any complex of semantic equivalents which exhibits such striking cross-cultural similarities. It is even more important to see that the roots involved very frequently imply an appeal to superior power for protection or redress against some sociohistorical power. This is the historical origin of prayer in the biblical meaning of the term.

28. "Vom Vindex zum Iudex."

29. I.e., so far as our sources go. The problem is that where the social institution existed we are not likely to have written documentation. Since our written sources are at present extremely minimal, the existence of something resembling the Roman Vindex is at present a prediction. We can only cite the Syrian *ekdikēs* as a possible candidate, but as a late survival of a no doubt *much* earlier social institution, or as a sociopolitical borrowing from an otherwise unknown Hellenistic political office. However, as a legal form of action, Koschaker has already pointed out the very close relationship between the *manus iniectio* of the Vindex in Roman law and the action termed *qātam nasāḫu* in old Babylonian law (P. Koschaker, *Babylonisch-assyrisches Bürgschaftsrecht* [Leipzig, 1911], para. 3, pp. 24–31). The traditions concerning the early Israelite "judges" may very probably, however, be relevant since there can be little doubt that the root *ŠPṬ* is an early synonym of *NQM*.

context sufficient to permit historical and legal analysis. There are only two groups of uses, both of them giving an appearance of stereotyped phrases (implying the probable archaic nature of the uses even by that time) and limited to certain types of historical contexts, as mentioned above.

The first group of uses originates in four cities of Palestine: Megiddo, Gezer, the Hebron area, plus one of unknown source. In *EA* 244:25–29 Biridiya of Megiddo says: *ù lu-ú-mi li-iq-qí-im-mi šarru^{ru} āla^{ki}-šu la-a-me yi-iṣ-bat-ši ^mLa-ab-a-ya,* "But let the king *rescue* his city lest Labayu seize it." He had earlier in the letter described Labayu's hostility toward him, which had the result of preventing normal activities, and was tantamount to a state of siege.[30] Four times in this letter he described Labayu's intention to "take" (*a-na la-qí-i*, l. 23), "seize" (*la-a-me yi-iṣ-baat-ši*, ll. 28, 37), or "destroy" (*a-ba-at-me*, l. 42) Megiddo. There can be no question of "avenging" here, and even Knudtzon regularly translates "*so rette . . .,*" though not in this case. What Biridiya requests in ll. 25–29 is the executive use of force in order to deliver the city from immediate peril. The specific executive act requested is then spelled out in lines 34–38: "Let the king give one hundred guard troops to protect [*a-na na-ṣa-ri*] the city, lest Labayu seize it." The two passages are perfectly parallel, and the second defines in specific detail the general meaning of the first. The root *NQM* signifies the executive exercise of power by the highest legitimate political authority for the protection of his own subjects.

Precisely the same usage is illustrated in *EA* 271:13, a letter to the king, from Milkilu of Gezer: *ù yi-qí-im šarru be-li māta-šu iš-tu qa-at^{lú.meš} SA. GAZ^{meš},* "So let the king my lord rescue his land from the hand [i.e., power] of the *SA.GAZ* [= ʿApiru]." The specification of the peril from which rescue is requested is designated by the preposition *ištu* with *qāt,* which is a precise analogy to the later Hebrew *nqm myd,* and thus demonstrates that both the grammatical construction and semantic range of meaning are closely related historically as well as linguistically. To anticipate, it is certain that both use and meaning of the verb remained constant from Amarna to King David; the vast difference consisted in the fact that the king of Egypt as the highest legitimate authority was rejected together with the administrative complex dependent upon him; instead, the God Yahweh now held that executive legitimate power, the actual exercise of which is designated *NQM.*

EA 274:10, from NIN.UR.MAH^{p1} of unknown origin, repeats Milkilu's words almost verbatim: *yi-qí-im šarru be-li māta-šu iš-tu qa-te^{lú.meš} SA.GAZ^{meš},* "Let the king my lord rescue his land from the hands of the *SA.GAZ.*" The fact that this differs from Milkilu's phrase only in substitu-

---

30. Ll. 9–17. For the context, see chap. V.

ting the plural of "hands" for the singular strongly supports the suggestion that the phrase is already formulaic—a scribal cliché, particularly since the two letters must be separated by some gap in both time and space. *EA* letter 274 seems to presuppose that Milkilu has disappeared from the scene since the closely related letter from the same person, 273, refers to the two sons of Milkilu.

The remaining three passages all come from Šuwardata, king of some city in the South, whether Hebron, Keilah, or other it is impossible to prove. *EA* 282:10–14 is again reminiscent of the passage from the letter of Biridiya of Megiddo: *yu-uš-ši-ra* ^{m}*šarri*^{ri} *bēli-ia ṣābē pí-ta-ti ma-aḫ-da danniš ù yi-qí-im-ni/ia-zi-ni*, "Let my king my lord send archer troops in great numbers that he may rescue me/deliver me." The gloss here helps to establish the significance. Unfortunately however, it is ambiguous, and has been taken from both *YṢ*^{ɔ} and *NS*^{c}. It now seems virtually certain that the latter is the proper analysis. The argument is as follows: the scribes of the Amarna letters probably thought they were writing good Babylonian in using the forms *yiqqim*, from Babylonian *ekēmu*, but were actually, as argued here, using it in a way which corresponded to West Semitic *NQM*. There certainly can be no doubt that there is a very considerable semantic overlap between the two words. The most common meaning of *ekēmu* is "to take away by force," but it is not used in the type of contexts found in the Amarna letters.[31] If the scribe wished, then, to give what he felt to be the proper West Semitic gloss to the word *yiqqim*, West Semitic root *NS*^{c} was actually the proper one to use. Several texts speak of *ekēmu* in reference to the removal of kingship; in Ugaritic *NS*^{c} is used in similar fashion: *lyṣ̌ alt ṯbtk lyhpk ksa mlkk*, "Verily he will snatch away the dais of your Residency; verily he will upset the throne of your kingship."[32] Compare also Psalm 80:9: *gefen mimmiṣraim tassī*^{ac}, "You did snatch away a vine from Egypt." In both cases, the idea "to remove by force" fits well the Babylonian usage of *ekēmu*, and may reasonably be assumed to have been the word which the scribes felt to be the appropriate synonym. The major difference between West Semitic *NQM* and Babylonian *ekēmu* lies in the context of usage, not in their dictionary definitions. The latter is almost never used of a legitimate exercise of force; in historical contexts it always indicates an unlawful force. The converse is true of *NQM*; it is never used except as an act of force carried out or authorized by the highest authority.[33] Further, *ekēmu* does not have any attested uses which could reasonably be translated as "rescue," as the Amarna

31. *CAD*, s.v.

32. I AB vi, 27–28. For the context, see *ANET*, p. 141.

33. This is true of the Amarna corpus, and with perhaps only a few exceptions, of the whole body of recorded uses. Note that it is an obligation under treaty and oath in the Sefire stela.

passages demand, though curiously enough both Babylonian *eṭēru* and Sumerian *KAR* do have a range of meaning which includes both uses.[34]

Šuwardata is of course not asking for physical removal from the land, but rescue from jeopardy; yet just as in the letter of Biridiya, it is the sending of soldiers in force which constitutes the means by which the king should perform the action requested: *yiqqimni*. It is an interesting fact which cannot be further exploited that only Šuwardata requests this action on behalf of himself personally. In all other letters of this type, it is the land or the city for which the action is requested.

EA 283:15–17, is not easy to interpret because of an unfortunate break in the tablet: *šum-ma mi-la an-na i-ia-nu ṣābē pí-ta-ti yi-iq-qí-mi-ni* ^m*šarri*^{ri} *bēli-ia*, "If there are no troops beyond these, then let the king my lord rescue me." The simplest interpretation is that Šuwardata claims that the forces at his disposal are not sufficient to defend himself against the thirty cities which he maintains are hostile to him, and that therefore direct action of the king himself is necessary. Later in the same letter the final illustration of this first group of uses is found (283:25–27): *yu-uš-ši-ra* ^m*šarri*^{ri} *bēli-ia ṣābē pí-ta-ti yi-iq-qí⟨im-⟩ni* ^m*šarri*^{ri} *bēli-ia*, "Let the king my lord send archer troops; let the king my lord rescue me."

The second group of uses consists of two passages in a single letter from *IM.UR.SAG* (which could be read Hadad-qarrad or Hadad-gabbar),

EA 250:20–22:

yi-qí-im-ni-mi ilim^{lim}
ša šarri bēli-ia
aš-šum i-pí-iš nu-kur-ti
i-na amēlūt ^{māt}*GI-na*
ardūt šarri bēli-ia

and EA 250:48–50:

yi-iq-qí-im-ni-me ilim^{lim}
ša šarri bēli-ia
aš-šum i-pí-iš nu-kur-ti
i-na šarri bēli-ia

Again, it seems clear that we are dealing with a scribal or at least conventional cliché which never recurs in later literature, though quite precise analogies are found elsewhere. Perhaps the most literal translation is also the most accurate: "May the gods of the king my lord rescue me from making war / against the people of the land of *GI-na* servants of the king my lord / against the king my lord." In the context of the historical situation, the statement is entirely comprehensible. Earlier in the letter, the sons of Labayu were quoted as saying to *IM.UR.SAG*: "Make war against the people of *GI-na*, because they killed our father; and if you do not declare war, then we will be your enemies." Lines 20–22 constitute the original response to the sons of Labayu. There is a rather delightful irony in the fact that the root *NQM* is

---

34. Friedrich Delitzsch, *Sumerisches Glossar*, and *CAD*, s.v.

here used as a ground for *refusing* to participate in a private war of vengeance, in view of the fact that biblical scholars have for decades assumed unquestioningly that *NQM* means primarily blood vengeance.

There is no immediate answer to the question of why appeal is made to the "gods of the king" rather than to the king himself. It is most tempting, however, to relate this to the fact that suzerainty treaties were in the Late Bronze Age sworn to by the vassal, naming the gods (whether of sovereign or vassal) as witnesses, i.e., enforcers of the treaty. The sender of this letter was indeed in a precarious situation, ordered to engage in an action which was clearly a breach of covenant under pain of suffering the consequences from clearly superior force. For him to appeal to those divine powers for protection against the consequences of acting in accordance with his sworn obligations is at least comprehensible. Unfortunately, the Amarna archives have given us little more than hints as to the nature of the Egyptian-Canaanite suzerainty treaties, though such certainly must have existed. The parallel passages in the same letter would tend to support this conclusion, for hostility against the "servants" of the king is later equated with hostility against the king himself. In all probability, this letter exhibits the historical foundations which are actually most characteristic of the new religious community a century and a half later. If it is not the king of Egypt, but the gods upon which his power depends, who constitute the ultimate ground of obedient action, and if it is not the people of the land of GI-na and not even the king of Egypt, but rather the very content of the obligation which takes precedence when a choice has to be made, then it follows that the primary problem for any unity and establishment of peace in the land must rest upon the widespread recognition of these two facts. These are actually the twin pillars upon which the entire Old Testament faith rested as a historical fact which, together with the community of human beings which constituted the twelve tribes (not one), make plausible its existence. Tribal religion was characteristic of the period before Moses and after Nehemiah, but not in the nearly thousand years between the two, in spite of the near success of the pagan tribalism of the monarchy.

As far as the form of the statement is concerned, at first glance it may seem to be an oath formula. Compare the very frequent *ʾaʿudu biʾllah* in Islamic literature, and the discussion in Lane, s.v., *ʿaḏa*: "One says also *maʿada ʾllah ʾan ʾafʿala kaḏa* for *min ʾan ʾafʿala kaḏa*. 'I seek preservation by God from my doing such a thing,' as though meaning 'May God preserve me from doing such a thing.' Qurʾan XII 79, and some reckon *maʿadu ʾllah* among the forms of oaths."[35]   A very closely related phrase is found in Isaiah 8:10: *wᵉyissᵉrēni milleket bᵉderek hāʿām hazzeh*, "And may he correct

35. E. W. Lane, *An Arabic-English Lexicon*, 8 vols. (London, 1863–85), p. 2192.

me from walking in the way of this people." It is clear that the Amarna idiom *aš-šum i-pí-iš* is a translation of Canaanite preposition *le* plus infinitive. Compare *EA* 136:30–32: *ù al-ka-ti a-na bīti-šu aš-šum e-pu-uš ṭābūti bi-ri[-nu]*, "I went to his house in order to make friendship between [us]." This expression of purpose clause introduced by *aššum* could not contrast morphologically to prohibitive clauses, since the Canaanite *le* was ambiguous, meaning both "to" and "from." So far as we now know it was only in later Canaanite that a grammatical distinction between the newly developed *min* and *le* was introduced.

From all of these examples in the Amarna letters, the conclusion that the root *NQM* designates what we have termed defensive vindication is inescapable.[36] Nowhere does *NQM* specifically imply anything similar to vengeance. Indeed, in the letters from Megiddo and Gezer the action is one which may be defined as the exercise of the imperium for the primary purpose of preserving it over lands which are under the attack of those who do not respect their obligations to the king. This violation of obligation consisted specifically of the illegitimate use of force, illegitimate alliances (which were of course tantamount to conspiracy), and other forms of harassment of neighboring states. Since the ability of the royal commissioners to maintain the peace by orderly means (no doubt theoretically available) had completely broken down, the various city regents under attack appealed to the imperial power directly. The exercise of that imperial power is called *NQM*; it is extralegal since legal process was ignored or rejected. It must be made clear, however, that from the point of view of political states, since the dawn of history, war and law have been considered two aspects of the same exercise of imperium, of legitimate authority. Both ultimately rest upon the actual possession of power to be exercised against the enemy, whether the internal criminal (law), or the external one (war). Those who reject or ignore the regular processes of law and social order can be dealt with only by force. It is such persons who are designated *ʿApiru/Ḫabiru* in most, but not all, of the occurrences of the word in the Amarna letters, including two of the letters cited above, numbers 271 and 274.[37]

Defensive Vindication in the Bible

Of seventy-eight passages in the Bible where the root *NQM* occurs, fifty-one involve situations in which the actor is either Yahweh Himself, or an agency to which the power to act is specifically delegated in a specific situation. Thus in over two thirds of the total occurrences, the root designates

36. Cf. also Latin *Jupiter Vindex*, and the *vindex* as the patron protector of a client (Düll, "Vom Vindex zum Iudex," pp. 112–13).
37. See chap. V.

the exercise of the divine imperium either directly or indirectly. Against whom is the imperium directed? Though there are a number of cases such as the stories of Cain and Lamech, where this particular issue is irrelevant, it is very significant that the exercise of the divine imperium is identified with an attack on a foreign population only in a bare majority of the uses, approximately fifty-five percent. Conversely, nearly twenty percent of the occurrences refer to the exercise of the divine imperium against Israel itself, or against individuals within the community. About one fourth of the passages are not relevant to this particular issue, since some refer to actions of foreign agencies, some are prohibitions of such actions to individuals within the community, and others are too generalized to be of value. Those uses of the root that are clear, however, strongly indicate that they have to do with the use of power against the enemies of that power, *whether internal or external.* In other words, the uses are analogous to the twofold use of royal power in law internally and in war externally. For this reason it is not possible at any time in biblical history to conclude that Yahweh is merely a tribal god of the ancient Israelites. Yahweh represents the legitimate power to which the religious community is absolutely obligated, and he is far from being merely a projection of "national self-interest" where this conflicts with an obligation that takes precedence as the interest of God himself.[38]

The foregoing discussion included all of the relevant biblical passages. Here the primary concern is with those few biblical occurrences of the root *NQM* that are analogous to the Amarna use of the word. Uncertainties of interpretation make it difficult to say in specific cases whether the action is defensive or punitive. After all, the rescue of one party from another in a conflict almost inevitably involves the use of force against the attacker, and therefore the defensive and the punitive aspects are merely two sides of the same coin. It is the time element, as is pointed out above, that determines the classification. A tort once done is not the same thing as jeopardy—the prediction that a tort will take place if superior force does not intervene.

The classical illustration of defensive vindication in the Bible is found in the confrontation of David and Saul, 1 Samuel 24:8-15, and the parallel account at 26:10-24. Unfortunately, the former account cannot be proven to represent tenth-century linguistic usage since it is usually assigned to the "late source" of Samuel. Yet the grammatical construction of the account has been identified as that characteristic of the Amarna letters which very strongly suggests that it does rest upon very old tradition. Furthermore,

---

38. The prophetic prediction of destruction of state and temple proceeds from this basic principle, of which the curses are of course the instrument. See Hillers, *Covenant*, chap. 6. It goes without saying that both Israel and Judah *attempted* to identify Yahweh with the respective "national interests."

there has not yet been sufficient progress in the history of the Hebrew language to justify uncritical reliance upon the traditional assignment of passages to early or later times. On the other hand, the very fact that verse 15 substitutes the root ŠPṬ 'judge' for NQM in verse 12 is good indication of the very close semantic similarity of the two.[39] This is particularly striking when one sees that ŠPṬ originally designated the action of a person holding an office, an imperium, in the Mari letters.[40] Also, as is well known, it became a political title in later Phoenician. Originally, then, the root probably had nothing to do with judicial action—like NQM, it indicated executive action.

Most important for the present thesis, however, is the fact that such an appeal to the executive defensive vindication of Yahweh excludes self-help on the part of David. Such actions in self-help constitute a claim to an imperium on the part of the individual which is incompatible with and actually rebellious against the Imperium of Yahweh.

An early example of defensive vindication is given in the famous episode of the battle of Aijalon, though the old poetic excerpt is woefully short of context.[41] Nevertheless, the historical narrative could have been taken directly from the Amarna letters as well as from later episodes of international power politics. The immediate result of the Gibeonites' covenant with early Israel was the mobilization of a coalition of kings to attack the defector rather than the main power.[42] A desperate call for help from Gibeon brought the assembled folk militia to defend the beleaguered ally-client, and the result was a victory for Yahweh. The brief poetic fragment celebrates the unusual meteorological phenomena which demoralized the enemy, and the result is described in the words, ʿad yiqqōm gōy ʾōyᵉvāw, "until he defeated the forces of his enemies."[43] It is a classical illustration of the thesis presented above, that the verbal root and derived nouns designate the use of force by legitimate sovereign authority, and that where such force is used in a situation involving armed attack, the usage of the verb may demand a translation into English by the word "defeat" or "rescue," depending on the context. If one views the situation as the hostility between sovereign and enemy, the

---

39. Note also the "Early Source" 26:24: "May he [i.e., Yahweh] deliver me from all distress," and the similar formula in response to an announcement of Yahweh's vindication in 2 Sam. 4:9: "... who has delivered my life from every jeopardy." Compare the reiterated formula in the "apologia" of Hattusilis: "Ištar, my lady saved me on every occasion" [... errettete mich bei jeder Gelegenheit]. See A. Goetze, "Hattusilis," MVAG 29 (1925): 11, ll. 43ff. for further detail, including the divine manifestation ḫandandatar as the agency of deliverance.

40. Cf. n. 24.

41. Josh. 10:12–13, from the "Book of Yashar."

42. Compare the Immanuel episode, Isa. 7:1–20, and numerous examples in the Amarna corpus.

43. Josh. 10:13.

word must be translated as "defeat" or "punish" (but compare below); if one views it as the relationship between sovereign and faithful subject, the same act is to "rescue" or "deliver." Only the context of the event can determine the proper translation, and the presence of jeopardy which culminates in armed conflict demands the translation "defeat." It is somewhat surprising that the Septuagint uses the Greek ᵓamunō, 'to ward off,' which is probably better than any word we have in English.[44]

If the defensive vindication of David discussed above illustrates this power in action in behalf of an individual, the Aijalon example illustrates the defensive vindication of the group. All other early examples have to do with military leaders: Jephthah, Saul—and, it is probable in view of David's peculiar status that his usage (discussed above) should be placed here also (cf. Šuwardata's peculiar usage above). It would be over-hasty to say that the root was never used to refer to the defensive vindication of persons by Yahweh in the early period, since, as it is pointed out below, such action on behalf of individuals is quite characteristic of Yahweh.

The story of Jephthah has the term under examination spoken by his daughter upon his successful return from armed conflict with the Ammonite king. "Do to me just as that which went forth from your mouth, since Yahweh has done for you vindication from your enemies" (Judg. 11:36). The form used here is the feminine plural noun with the common verb "to make or do." This same feminine plural occurs also in the early Psalm 94:1: ᵓēl n^eqāmōt hôfīᵃᶜ, and in the Davidic Psalm 18:48: hāᵓēl hannōtēn n^eqāmōt lī, "the God who gives Imperium to me."[45] It should be regarded as a plural of totality in all probability, in the context of the delegation of power by God to the king, and therefore designates the totality of the legitimate executive force. With this passage, the use of the root virtually ceases in biblical Hebrew for more than three hundred years, with only a poetic archaism occurring in Isaiah 1:24 and Micah 5:14,[46] chosen because it alliterates so beautifully: ᵓennāhēm/$w^eᵓ$innāq^{e}mā, and in the similarly archaic reminiscence of Deuteronomy 32:43 used by the prophetic party at the anointing of Jehu to overthrow the bloody dynasty of Ahab and Jezebel: "that I may vindicate the blood of my servants the prophets" (2 Kings 9:7).

Two occurrences of defensive/punitive vindication in early sources date from the reign of Saul. The first of these (1 Sam. 14:24) has to do with the oath of Saul and the inadvertent violation of it by Jonathan: "Cursed be

44. The Greek word also has overtones of 'avenge'—to inflict harm in retaliation for a past injury, which is irrelevant to the context of this narrative.

45. There is no other similar construction in early uses except in 2 Sam. 4:8, where it is definitely vindication from the house of Saul.

46. Both passages belong to the category of punitive vindication, for which see below.

the man who eats food before evening, and I am rescued from my enemies." The entire episode probably illustrates the nonfunctional, superstitious religious concepts of Saul, who seems to think that self-affliction through ritual behavior and deprivation is a guarantee of divine support, just as later, in the case of the battle against Amalek, he seems to think that ritual feasting is an adequate substitute for obedience to divine command (15:15), and is bitterly upbraided by Samuel. Unfortunately, Saul's ideas of religion seem to have been institutionalized in the utterly pagan Temple of Solomon, as the prophets never tired of pointing out.[47]

The second usage from the time of Saul occurs at 1 Samuel 18:25, and adds no further information, except possibly the fact that Saul is using the old religious concept that Yahweh exercises His imperium through the acts of individuals, for the purpose and in the hope that David will meet death in the process of gathering a hundred Philistine foreskins. If so, it well illustrates the degeneration and cynicism which attended the establishment of the monarchy.[48]

The same is even more true of a last early use of the root. In the most interesting and informative passage at 2 Samuel 4:1–12, two disreputable turncoats commit political assassination in order to ingratiate themselves with the man who will inevitably win power. They piously represent theirs as the act of God: "And Yahweh has given to my lord the king deliverance [i.e., $n^e q\bar{a}m\bar{o}t$] this day from Saul and from his dynasty." David, in his ensuing oath, affirms that Yahweh has indeed delivered his life from all danger, but to reward evil men for murdering an innocent man in his bed is out of the question, and they must be extirpated. According to the old principle that he who protects a murderer is himself guilty of the same crime and subject to the same punishment, David had to disavow any complicity and reject any suspicion of even having condoned the crime by putting the assassins to a shameful death.[49] One can easily see the dilemma faced by David: Joab, his most faithful ally, relative, and support, had shortly before been guilty of the same crime of political murder, motivated no doubt by the same consideration—hope of enhanced political status by removing political rivals or obstacles. It is this conflict of loyalties that led David to have been satisfied with a curse upon the house of Joab—he was willing for personal reasons to take the risk of letting Joab live, but did not wish the risk to fall upon his

47. In view of the narratives we have, the ritual substitutes of the sacrificial system of the temple were absolutely necessary. A strict adherence to the old Mosaic religious law would probably have left virtually none of the royal family alive.

48. There is a striking similarity to the Amarna usage of Šuwardata in these royal uses, in that it is always the person of the king, not the land or the people, that is the object of divine vindication. The other usage is illustrated in Deut. 32.

49. See chap. IV for the exposure of the corpse after execution.

descendants.[50] This accounts for the last instructions of David to Solomon to execute the death penalty for murder (1 Kings 2:5). Yet David admits that he had put to death a man simply for bringing what he thought to be the "good news" of the death of Saul and Jonathan. Do we already have here the sorry old idea that the king (i.e., the state) can do no wrong?[51]

With the establishment of the state and the delegation of the divine authority to the monopoly of force, the ancient concept of the Imperium of God and the use of the word which designated it came virtually to an end. As in other ancient pagan states, the very idea that one could appeal to God against the decision or acts of the one to whom God had delegated final authority was a logical and political absurdity, which only the prophetic movement persisted in[52] (particularly in the north after the division). Unfortunately, a large segment of the population subject to the state (which since David included much of the old Canaanite urban populations) found the state intolerable, and had no other recourse than God or some other god to whom to appeal. One result was a series of solemn curses on the state and, particularly in the south, the pronouncement of destruction of the monopoly of force and its established Temple, whether northern or southern (ironically enough, in ancient terminology, leftist or rightist).[53]

Then, shortly before the political end, a reexamination and revaluation of the old traditions began, but it was too little and too late to avert the horrible catastrophe. It was under these circumstances that the concept of the "vengeance of God" found expression in forms of language which have become empty threats in the English language.[54] New linguistic forms and structures, having only remote connection with the old forms and ideas, begin to occur in relatively great number during the chaotic and tragic period from Jeremiah to Deutero-Isaiah, from about 600 to 540 B.C.

50. Compare the words of the woman of Tekoa (2 Sam. 14:9). To protect the guilty is to share the guilt—and the ultimate divine punishment. By this formula, she exempts the king and his descendants from such participation.

51. 2 Sam. 4:10. The problem in the narrative of 2 Sam. 1 has often been noted. The fact that David does not refer to the Amalekite's putting Saul to death (albeit upon request) strongly suggests that 2 Sam. 1 exhibits an attempt to place David's action in a better light, especially since he had just returned from a punitive raid on the Amalekites.

52. It was a short step, already made by Amos, and probably foreshadowed by Elijah, to realize that it is futile to condemn the king for policies and acts that represent the popular consensus. If there were only 7,000 who had not bowed the knee to Baal, the entire community would have to share the blame—and to share the catastrophe that took place in 722–21 B.C.

53. See chap. I, n. 112. Zephaniah almost certainly falls within the context of Josiah's reform, when the expectation of a real reversal of the march toward doom was quite alive—but short-lived!

54. The concept of the punitive *deus vindex* was still alive in American legal practice as late as the end of the nineteenth century. Cf. chap. VIII, n. 9.

Ever since Amos, the most able representatives of the old religious/ revolutionary tradition whom we now know as the "canonical prophets" had seen the simple fact that it is quite foolish to blame the government for social ills which derive from the people governed. If politics is the art of the possible, why blame the government if those governed will not tolerate anything but the usual delicate balance between hogs, rats, and weasels? Is all mankind to be divided in a tripartite fashion between those who are weighty enough to get both feet into the feed trough, those who are stopped seemingly by nothing to get what they want, and those who destroy and kill just for the fun of it? It is within this context of historical reality that the prophetic crescendo of impending doom must be understood. To understand the religious foundation of the confident predictions, we must turn back to the older traditions of the formative period and the word uses which illustrate the "punitive vindication" as a kind of recognized exercise of the divine imperium. Even though, with the exception of Isaiah 1:24 and Micah 5:14, the earlier prophets never use the word, their entire message is a continuity of the earliest religious ideas and convictions.

PUNITIVE VINDICATION

Apart from the single published occurrence of the root in the Mari documents mentioned above, we have no prebiblical illustration of punitive vindication as the meaning of the term *NQM*. It is entirely possible and even probable that the two occurrences in chapter 4 of Genesis are a survival of very old usage. Yet even in these passages, which have to do with non-Yahwistic, and perhaps even non-Semitic, cultures, the custom of blood vengeance as in the case of Cain, or that of unlimited retaliation or reprisal for injury as in the case of Lamech, have religious overtones. In the first place, both passages use the *Qal* passive, just as in the Covenant Code (Exod. 21:22) or the *Nifʿal* (21:20). Though this can hardly constitute absolute proof that the logical subject of the action is Yahweh, the narrative which attributes the custom to Yahweh's intervention and authority very strongly implies as much. Furthermore, we find precisely the same structure of thought in usages of the same root in pre-Islamic Arabic inscriptions, in which appeal is made to a god to make available redress for a wrong (*NQM*) or blood vengeance for the death of a relative.[55] It is of particular importance, however, to note that *NQM* never seems specifically to refer to blood vengeance; for this another root *T̠ʾR* (cognate to Hebrew *ŠʾR*, 'flesh'?) is used, as also frequently in classical Arabic (see Lane, *Arabic-English Lexicon*, *s.v.*).

55. See above, n. 17.

It is probable, furthermore, that the narrative of Cain actually had an important historical-political function originally. Coming as it does from the "J Document," or whatever other designation the literary critic may wish to assign to it, it is almost certainly from the period when the Kenites not only maintained a corporate identity but also were incorporated into the political dominion of David and Solomon.[56] The narrative that places Cain under the protection of Yahweh would then be a religious and political recognition of the custom of blood vengeance, which would be especially important to a group that traveled long distances among a variety of cultures as a means of helping to secure their freedom from attack.[57] The same political adjustment to reality is illustrated in much of the legal history of the Near East—even if central governments would prefer to eradicate the custom of blood revenge in favor of normal legal procedure. By placing the tribal custom under the protection of Yahweh, the official religious and political establishment thus permitted the continuation of the tribe and avoided unnecessary conflict. Regardless, the tribe disappeared before long. However, it is possible that this illustrates the old custom of identifying the respective gods of two groups for political reasons.[58]

The specifically Yahwistic uses of the term in the old sources prior to the Divided Monarchy consist of some ten occurrences, which, again, fall naturally into two categories, depending upon the relationships involved in the case. If it is the legitimate function of the sovereign to use force to repel illicit challenges to his authority, these challenges then may involve either actions against the authority itself, or actions consisting of past attacks on those under the protection of the sovereign. Such categories are now classified either as rebellion (if from within) or aggressive attack (if from without), or, in the second category, punitive action against those who

56. This is indicated by the "genealogical list" of 1 Chron. 2:55, among other grounds. Note also the participation of Kenite cities in the spoils sent by David, 1 Sam. 30:29.

57. The reference to the "cities" of the Kenites should serve as a warning not to equate a social group designation with a particular socioeconomic adaption to environment such as nomadism. The probably non-Semitic origin of Cain/Kenites is indicated first by their close association with the Midianites (see chap. VI), and by the name. There is a curious parallel between the semantic development of the roots $\underline{T}^{\circ}R$ 'kin' (Ugaritic and Hebrew) and Arab $\underline{T}^{\circ}R$ 'blood vengeance' on the one hand, and the root kaena- 'Verschwägerter, Verwandter' in Hittite, and Persian kaen- 'Blood Vengeance' on the other. It may be coincidence, but a similar root, kan- in Persian means 'digger'— i.e., miner—and we have thus the two cultural traits attributed to the eponymous ancestor whose home was originally in Tubal, Anatolian Tabal, well known as an important metallurgical center in ancient times.

58. Thus hypothesis of the Kenite origin of the name Yahweh is completely without foundation unless there is other compelling evidence free of the suspicion of such political accommodation.

have attacked those under the protection of the sovereign, action that is categorized as crime or war. The early uses which indicate punitive vindication include both types with their subdivisions.

First, there are the Covenant Code passages often cited to demonstrate that blood vengeance was a normal social custom in early Israel. Only such cases as the death of a "slave"[59] at the hands of his master are involved. The normal procedure in the case of murder seems to be as follows: the murderer flees to a "city of refuge" by which means his life is not only protected, but any private person who may feel justified in taking the law into his own hands is also thereby prevented from engaging in such action. The case is tried before a normal village court, and if the defendant is found guilty, he is then turned over to the *gōʾēl haddām* ('redeemer of blood') for the execution of the punishment. This *gōʾēl haddām* seems usually to have been the next of kin, who, in the absence of any public executioner, acts as the agent of the court for the execution of the verdict.

The slave, however, presumably has no one to act as the *gōʾēl haddām*,[60] and probably no one who will convene the court or make the necessary indictment of the murderer. As jurists never tire of saying, the law is not self-enforcing. It is because of these special circumstances of the cases involving persons who have the technical-legal status of slaves, that the term *NQM* is used in the oldest law collection of the Bible. In such cases, the executive authority of Yahweh Himself is the basis for community action against the slave-owner, but only after the necessary court inquiry. The law makes a very commonsense distinction between cases in which the slave dies under the hand of the master, and those in which the slave dies some time later as a result of a beating. The presumption is that a slave-owner has an interest in preserving the life of his own property, and there was therefore no guilty intent to murder unless the slave died immediately. The action *yuqqam* presumes, then, that the community itself is charged with exercising the executive authority of Yahweh to punish the murderer, and thus at the same

59. The term "slave" probably should not be used to translate the Hebrew *ʿeved* in sources that stem from the period between Moses and David. Although they were evidently regarded as property, the "slave" status must always have been a voluntary one, for the society had no social machinery by which to compel the return of a runaway, and the evidence of somewhat later law shows that such a return was prohibited. The Covenant Code demonstrates rather that "slaves" were really indentured servants, and it took an act in the presence of God (i.e., as witness?) to create a servitude in perpetuity. The act therefore partakes of the nature of a solemn covenant which became irrevocable, as the pierced ear testifies. Further, it is the solemn act and deposition of the "slave" himself, not the master, and we even have the historical prologue: the gift of wife and children, which furnishes the motivation.

60. According to Exod. 21:1–6, the slave is an *ʿApiru*, i.e., a person without a legal community.

time giving (posthumous) redress to the murdered slave. Since the passage is ambiguous—that is, it is impossible to say whether it is the slave or the master who is the subject of the passive verb and therefore the object of the action— the question of whether or not the verb is actually impersonal immediately arises. In other words, the passive verb has neither slave nor master as the logical subject. Rather, it is a command that the sovereign authority of Yahweh should be placed in action in order to punish/redress an action that is incompatible with the sovereignty of that same ultimate authority. Under the covenant, for the community to fail to take action against a murderer is for all to assume the burden of guilt for conspiring at and condoning an act of murder. All, therefore, risk the consequences of the wrath of God under the curse formulas of the covenant. Again, we can see that the ethic of the new religious community transcends the tendency of tribal groups to protect members who are guilty of gross delicts. The most dramatic illustration of this operational principle of ethic is to be found in the narrative concerning the rape of the Levite's concubine (Judg. 19–20).[61]

Though it has doubtless always been difficult for a community to punish a substantial citizen for the murder of a slave (ancient Rome gave up entirely such attempts), there is no reason to doubt that a law such as that described above was actually a part of real legal practice. The motivation is not far to seek. In the first place, on the social-historical level, murder that goes un- punished calls into question the security of every person in (and outside) the community, and elicits horror among any but the most calloused of com- munities. This is illustrated even in ancient prebiblical texts, even in cases in which the execution is carried out by legitimate political authority.[62] It is expressed in the biblical passages which represent the blood of the murdered as crying out to God for punitive action (Gen. 4:10).

The punishment associated with divine action, however, characteristic- ally falls upon the entire community. A most dramatic illustration of this idea is provided by the Aqhat Epic from ancient Canaanite Ugarit; in it, unpunished murder results in a seven-year drought which is of course catastrophic for the entire community.[63] In early Greek sources the same concept causes periodic human sacrifice to the Minotaur.[64] Finally, the

61. The same narrative supports the conclusion reached above with regard to the status of "slaves." The Levite had to go in person to the "slave" girl to persuade her to return to him. Use of force is not even hinted (Judg. 19:3).

62. For example, the report to the king of finding a dead child in the street at Mari (ARM(T), VI: 43. The vigorous protest and accusations against Rib-Addi for the execu- tion of those probably guilty of sedition (EA 138:39ff) illustrates the same reaction.

63. Aqhat C, col. i. ANET, p. 153.

64. Diodorus IV. 61. The story was certainly already known in the eighth century B.C. (PRE minotauris, col. 1929).

same procedure of curse and drought is a consequence of unpunished murder carried out by the state itself in the Elijah cycle of narratives.[65] In ancient Israel, however, the social punishment is carried out not upon innocent victims, but upon those who collectively were responsible for the deeds, who systematically defended and adhered to the policies which supposedly justified the murder of the innocent—Jezebel and the house of Ahab.[66]

This situation within which the verb NQM is used in 2 Kings 9:7 is the only usage of the verb in a narrative context between the time of David and the days of Jeremiah. At the command of Elisha, a young member of the prophetic order proclaims the message of Yahweh: "I anoint you king ... and you shall strike down the house of Ahab, that I may redress the blood of my servants the prophets from the hand of Jezebel." There can be little question that the phrase, whether its origin is in the event itself or in later interpretation of the event, is a reminiscence or actual quotation from the old pre-Monarchic hymn, Deuteronomy 32:42. Again, it is clear that the action designated as the vindication of Yahweh takes place in a context which makes any normal legal procedure for redress absolutely impossible —for the action is called for against the state itself and its agents. Nothing could better illustrate the survival of the old concept that Yahweh is sovereign over the state and must act eventually (though it be only in the third generation as in this case) to redress the murder of the innocent, especially when it was done for political reasons.

The final passage in which NQM has to do with punitive vindication in redress of wrong suffered by an individual is the story of Samson. This thoroughly atypical narrative yields to no easy interpretation. It has been suggested that in essentials it is a narrative of actual events that was incorporated into the biblical narrative partly because it was associated with the half-century long struggle against Philistine occupation, and partly because it was such a parade example of the futility of the escalation of wrongs, reprisals, and counterreprisals. It is told objectively with no obvious attempt to whitewash Samson or to make the Philistines appear in as evil a light as possible. Particularly note the almost eager attempt on the part of the

---

65. 1 Kings 17:1. Though the narrative as we have it does not specifically connect the curse with Jezebel's judicial murder of the prophets, there can be little doubt that this was actually the ground (1 Kings 18:4 and 2 Kings 9:7).

66. The collective and indiscriminate nature of *divine* punishment of an entire community for the sins of segments or even individuals was of course widely known in the ancient world. Cf. Pindar *Pythiae* III. 34–37, where Artemis, the punitive emissary of Apollo, slays the adulterous Coronis, and "many of her neighbors suffered the same fate, and perished with her; even as, on a mountain, the fire that hath been sped by a single spark layeth low a mighty forest" (E. J. Edelstein and L. Edelstein, *Asclepius* [Baltimore, 1945], p. 4).

Philistine father-in-law to give Samson the younger "more beautiful" daughter, since he had mistakenly thought Samson had divorced (i.e., "hated") his wife (Judg. 15:2).[67]

In this complex of narratives the verb occurs twice—once when Samson vows to "be vindicated" because the Philistines had burned his father-in-law and wife as a punishment for Samson's reprisal against a whole group because of the (perhaps honest) mistake of one person.[68] Again, this is an individual action in a context in which legal procedures were either impossible or were rejected because of the nonexistence of a competent court of law. The entire narrative as we have it strongly insists upon Samson's status as an agent of Yahweh, both in the birth narratives and in the "spirit of the Lord," which is associated with his status as a Nazir. This is reinforced by his final petition to obtain redress by his own action for the loss of his eyes, the *Nif^{c}al* form (as also in Judg. 15:7), which seems most probably to be a passive rather than a reflexive usage (Judg. 16:28).

The close similarity of usage between the Samson occurrences and the language of Saul at 1 Samuel 14:24 and 18:25 (all *Nif$^{\circ}$als*) strongly suggests that kingship ideology is involved, especially since the root is also involved in the Jephthah episode (cf. Judg. 11:36) in the same form in which it reappears in the time of David: "The God who give *n^{e}qāmōt* to me . . ." (Ps. 18:38 = 2 Sam. 22:48). It is very tempting to conclude that in the development and gradual assimilation to the old paganism of pre-Israelite ideology, the concept of Yahweh's granting individual acts of successful exercise of power over enemies in specific situations of jeopardy had finally traveled full circle to end in the age-old political ideology that Yahweh gives a totality of power over enemies to the king. Corresponding to this is the fact that though the earlier wars were most certainly defensive (especially Aijalon), in the time of Saul they were fought for independence from Philistine domination. In the early reign of David the wars continued to successful completion, and were immediately followed up by aggression against nearly all border states.

With the disaster of Solomon's reign and the accession of his son whose ideology went far beyond his ability to translate it into reality,[69] the political

67. The law which prohibited the "best man" ever from marrying a (rejected) bride may be relevant. What is more important, however, is the fact that the father of the girl evidently did not expect a formal legal act of divorce or breaking of the betrothal. He could hardly have been held responsible if no divorce law recognized by both parties existed—as is probably the case.

68. The reprisal of the Philistines was not against Samson or his village, but against those responsible for Samson's act. Even the Philistines could recognize the fact that a wrong had been done to an outsider.

69. 1 Kings 12:14.

usage of the root *NQM* came to an end. It continued only as a designation of the old expectation of divine intervention in human history, but as pointed out above, in a radically changed context—historically, linguistically, and theologically.

To these discussions of early usages of the root which combine to demonstrate reference to the exercise of force in an extralegal executive context by a person or community to whom power to act is delegated by the one who holds total legitimacy of power, or an executive action by that power—Yahweh Himself, we may add one very interesting passage in which we have a similar situation, even a similar English translation, but in which the root under consideration does not occur.

The narrative in question is the story of David and Nabal (1 Sam. 25). David's claim to a "gift" at sheep-shearing time is clearly based upon the fact that his band of fugitives from King Saul (cf. 1 Sam. 22:2) had protected Nabal's flocks and shepherds in the wilderness: a prior and gracious act entirely analogous to the content of the historical prologue in Late Bronze Age covenants. Having benefited from such a gift, according to the old principle of reciprocity, Nabal was placed under obligation to David and his band, even though there had been no prior covenant. Nabal, described as a miserly skinflint, piously took refuge in the worst sort of self-righteous political legalism in order to evade his human and moral responsibility. His words may well be taken as classical illustration of the usual attitude toward bands of ʿApiru: "Who is David? Who is the son of Jesse? There are many slaves nowadays who are defecting from their masters." The implication, in other words, is that David has no social or political status from which to make any claim upon a substantial citizen. Furthermore, he is a "fugitive slave"—the term translated "fugitive" occurs nowhere else in the Hebrew Bible in this form and corresponds semantically (though not formally) to Amarna *paṭāru*, 'defect.'[70] The situation described is that underlying the cultic confession of Deuteronomy 26:5, where we should read, "A fugitive Aramean. . . ."[71]

Infuriated by Nabal's refusal to act in accordance with the elementary principles of reciprocity, David resolves to destroy him. Correctly predicting the consequences of such high-handed action, Abigail immediately acts without her husband's knowledge or consent to correct the wrong. Of primary interest here is David's response. After Abigail's long speech concerned with dissuading David from engaging in violence to redress a wrong, he replies with the same words she has twice used (25:26, 31): "Blessed be you who have restrained me today from blood guilt, and *saving* myself with

70. *EA* 106:20; 114:21–2; 118:37–9.
71. This was pointed out by Albright many years ago. Cf. chap. VI, n. 76.

my own hand." The root *NQM* does *not* occur either here or in the entire narrative. Private vengeance for a wrong is in Hebrew as in English, "self-help" (Hebrew *YŠ͑*, which is the usual theological term for "save," "salvation," "Savior").

To summarize, though David and his band were an ͑*Apiru* group (to use the Amarna term), i.e., they had neither status in, nor protection from, a political state, and though they existed in a most precarious condition in which they could continue to survive only through self-help, no form derived from the root *NQM* is used in the entire narrative. It is a term which can be used—or at least was used in our sources—only in connection with the legitimate exercise of force which in early biblical sources must have some socially recognized authority derived from Yahweh. All exercises of such force by individuals are hostile acts: compare Psalm 8:2. Only one exception to this rule exists in biblical literature, and it occurs in the book of Proverbs, which is notoriously illiterate theologically so far as its early materials are concerned. It has to do, however, with the situation in which a deceived husband engages in self-help to punish the paramour (Prov. 6:34), a privileged act in many societies.

With reference to the early usages of *NQM*, one must conclude that the normative value system of the early biblical society would never tolerate an individual's resorting to force in order to obtain redress for a wrong suffered. Perhaps better, such usage is not found in what has survived in biblical language. On the contrary, the very many references to Yahweh as the one who acts with legitimate power, which is most characteristic of all usages in the Hebrew Bible, powerfully reinforces the main thesis of these essays that in the thought of early Israel Yahweh had actually succeeded to and replaced the kings and empires of the Late Bronze Age as far as His community was concerned. Yahweh was the sovereign to whom alone belonged the monopoly of force.[72] Self-help of individuals or even of the society without authorization of Yahweh was an attack upon God himself:

Ah, I will vent my wrath on my enemies,
and obtain redress from my foes. (Isa. 1:24)

VINDICATIO IN THE LATE SOURCES

It has been observed above that during the whole period from Solomon to Jeremiah there are only three occurrences of the root *NQM* in datable contexts, two of which occur in the prophets of the south, Isaiah and Micah,

---

72. Exod. 15:18.

and one of which occurs in the prophetic message to Jehu, the only occurrence in the context of historical narrative. With Jeremiah, however, an enormous revival of usage occurs, which reflects, first of all, the impending breakdown of the political order and increasing appeal to extralegal power. Second, the destruction of state and temple organization brought about a most profound reexamination of the old traditions, a highly creative readaptation, and the reintroduction of many old and obsolescent religious concepts into the mainstream of community thought.

The Book of Jeremiah actually contains two clusters of usages in radically different contexts. The first has to do exclusively with internal conflict, and both the defensive and punitive aspects of *NQM* are involved. The second cluster occurs in the oracles against foreign nations, and is exclusively external-punitive in its references, which, with one exception, occur after the destruction of Jerusalem. Without exception, the references show the phenomenon of identification which we have already seen in the case of warfare in early times: namely, that a war is identified as the exercise of the sovereignty of Yahweh, and the defeat of an enemy is thus termed the punitive vindication of God. Thus, at Jeremiah 46:10, Nebuchadnezzar's defeat of Necho is identified with Yahweh's obtaining redress from his enemies, and the source of phraseology used is Deuteronomy 32:41, 43, but with different construction. All the other occurrences of *NQM* in chapters 50 and 51 have to do with the anticipated destruction of Babylon by the Medes (which never took place), and again the destruction is simultaneously the punitive redress obtained by the attackers and the retribution of Yahweh (50:15). In two passages, 50:28 and 51:11, the destruction is identified with the *niqmat hēkālō*—the "vindication of his temple," which is a curious and unparalleled usage,[73] both in prose additions to the poetic oracles and almost certainly in later additions.

In this collection of oracles against Babylon, three out of the eight uses that may be termed "attributive" are also to be found. The grammatical constructions involve the construct state *n^eqam-* or *niqmat-*, followed either by a personal pronoun (suffix), or by another noun which indicates the beneficiary of the action, whether defensive or in redress (punitive). It may be regarded also as an objectification of the action, which, in ancient times, as we have seen, is the exercise of the Imperium of God. In late pre-Exilic and Exilic times, there is a subtle shift in usage. Compare Leviticus 26:25, where the catalogue of curses includes the punitive sword: *nōqemet n^eqam b^erît*, "redressing the redress of the covenant" (or some such translation). Similarly, in Jeremiah 50:28 and 51:11, the *niqmat-Yahweh* is identified with the *niqmat-hēkālō*, "the redress of His temple." The old formula in

73. Somewhat similar is the Holiness Code phrase *nōqemet b^erît* (Lev. 26:25).

Deuteronomy 32:43 is similarly objectified in Psalm 79:10, *niqmat-dam-ʿavādāw*, so that it is no longer primarily the executive action of authority which is the determinant of usage, but rather the desired redress of bloodshed. In some such way, the transition from vindication to vengeance can be seen in operation.

This shift may already have been characteristic of the secular, paganized usage in the time of Jeremiah. In the confrontation with his enemies (chapter 20), they say (verse 10): *niqqᵉhā niqmātēnū mimmennū*, "Let us obtain our redress from him," but since the implication is that the procedure is entirely one of self-help, there is very little contrast to the modern action which is called "vengeance." There is, however, a sharp contrast to the response of Jeremiah in verse 12: *ʾerʾeh niqmātᵉkā mēhem, kî ʾēlekā gillîtî ʾet rîvî*, "Let me see thy deliverance from them, for to you I have entrusted my case." Jeremiah's response is entirely analogous to the uses of the Amarna period, for it is deliverance from jeopardy of life that he requests, not merely retaliation for a wrong suffered—the wrong is still in the future—and from which he barely escaped.[74]

The case is even clearer in the parallel passages, especially Jeremiah 11:20, where appeal is made to Yahweh as the "righteous judge" to deliver the helpless lamb being led to the slaughter (verse 19) from the conspiracy intended to "cut him off from the land of the living" (verse 19). The prose sequel gives the divine verdict: the punishment of the conspirators who threatened the death penalty for the messenger (prophet) of Yahweh is their own death by sword and famine, which is the "year of their visitation" (*pᵉquddātām*). Never, however, does Jeremiah speak of *my nᵉqāmāh*, as do his opponents. It is always an action of God, or it is an illegitimate claim to sovereignty which constitutes a challenge to that of God Himself.

It is not surprising, then, to find at Jeremiah 15:15 the very complex construction *hinnāqem lî mērōdᵉfai*, "obtain satisfaction for yourself for my sake from my persecutors."[75] Actually the passage is not translatable. The sequel, however, makes the situation clear. The patient forbearance of Yahweh is likely to result in the death of the one who is in jeopardy precisely because he did obey the command of the sovereign. Again, the similarity to the Amarna period is striking. As in an old Roman proverb, "In the evil society, the good man is the criminal." But the recognition of evil is easily seen to be the resort to force, slander, and conspiracy to silence one who rejects the resort to force.

In addition to the confrontation described in the three passages discussed above, there are three virtually identical refrains in Jeremiah 5:9, 29,

74. Jer. 26.
75. John Bright, *Jeremiah*, in *The Anchor Bible* (Garden City, 1965).

and 9:8 which use the *hithpaʾēl* form: "For these [sins] shall I not bring retribution? Upon a *gōy* like this, shall not my soul obtain satisfaction?" Elsewhere, only the two passages in Psalms 8:3 and 44:17 use the *hithpaʾel* form, and both refer to the illegitimate resort to force identified with the "enemy." It is in fact very tempting to conclude that Jeremiah is deliberately using rather shocking language: if his people act like the *gōyîm*, the heathen, then Yahweh will have to act thus also. Jeremiah is also fond of using the root *PQD* in association with *NQM*. This goes back to the ancient Decalogue usage, which affirms that Yahweh is a zealous, fervent God who does not indefinitely tolerate a flagrant violation of his commands, but will "visit"[76] the iniquities upon the lineage of those who have broken the covenant relationship with him by violation of his commands (= "who hate me").

To summarize the uses of *NQM* in Jeremiah, we may well quote the English dictionary: "AVENGE, REVENGE both mean to inflict pain or harm in return for pain or harm inflicted on oneself or those persons or causes to which one feels loyalty. The two words were formerly interchangeable, but have been differentiated until they now convey widely diverse ideas. AVENGE is now restricted to inflicting punishment as an act of retributive justice or as a vindication of propriety: 'to avenge a murder by bringing the criminal to trial.' REVENGE implies inflicting pain or harm to retaliate for real or fancied wrongs: 'Iago wished to revenge himself upon Othello.'"

The very vivid situations illustrated in Jeremiah, then, indicate that only in the passage where he quotes his enemies is "revenge" an at all proper translation (plus the two Psalms in which the revenger of one's self is equated with the "enemy," who is caused to cease by the power of Yahweh). "Avenge" does not fit, since the root in question often refers to situations where no wrong has as yet been done, and the action requested of Yahweh is one which transcends normal legal procedures established by society, even though legal language is used in the appeal to God. Always, it is the legitimate sovereignty of God which is the starting point for the word usage: his power to act, which is then often "identified" with historical events, particularly in war, but also in calamities past or future that, under the covenant, are punitive executive actions of God that must take place if God is to be not merely a symbol of the usual heathen interests of the group: power and wealth.

FROM THE EXILE TO THE NEW TESTAMENT

The enormous calamity of the destruction of Jerusalem is reflected in Exilic uses of the root, in addition to those already discussed in chapters 50–51 of

76. I.e., 'impose the consequences of. . . .'

Jeremiah. Ezekiel reiterates the message which Jeremiah also pronounced: the punitive vindication of Yahweh is inevitable. In the savage prophetic condemnation (Ezek. 24:2–14), dated the very day the king of Babylon laid siege on Jerusalem, the Babylonian attack is "identified" with the wrath and *Nāqām* of Yahweh, exactly as later (and earlier, cf. Jer. 46:10) other armies act as divine agents to punish Egypt, Babylon, and Philistia.

In Ezekiel 25:12–14 and 25:15–17 there are two highly stereotyped predictions of doom upon Edom and Philistines, which add little new except a peculiar construction that throws considerable light upon chapter 31 of Numbers. In 25:14, Ezekiel predicts Yahweh's action as a punishment for Edom's *ᶜᵃśōt . . . binᵉqōm nāqām lᵉvêt yᵉhūdāh*, "acting with punitive force against the house of Judah," and the sequel is: *wᵉnatattī ʾet-niqmātī beʾedōm bᵉyad ᶜammī yisrāʾēl*, "and I will put my *nᵉqāmāh* on Edom by the agency of my people Israel." The similarity in grammatical construction (which is very characteristic of Ezekiel—the cognate accusative in verse 14 and the phrase *nātan nᵉqāmāh bᵉ-* are both Exilic constructions) to the two repetitions in Num. 31:2–3 strongly suggests that Ezekiel is the source for the priestly (P) language in this narrative, which is probably an authentic tradition, though the language is late. Actually, the historical situation in chapter 31 of Numbers is also one in which there is jeopardy, the expectation of an attack as a result of the unilateral action of Phineas in murdering a Midianite princess out of misplaced zeal.[77] The consequences lasted a long time, and the Gideon story is the next sequel.[78] It would seem most probable that Ezekiel's mention of the "agency [lit. 'hand'] of my people Israel" actually rests upon a fancied identification of Edom with the Midianites—for which there may well be some basis in historical fact. In contrast, there is no agent mentioned in the prediction of doom upon the Philistines at verses 15–17.

In chapters 34–35, 40–66 of Isaiah, there are six occurrences of the root which seem to add something new. Three of the passages are what might be termed highly "time-bound": the vindication is associated with a time point of expectation either for punitive action against Edom (34:8, 63:4), or Babylon (47:3), or, on the other hand, an action which is not so much hostility against enemies for the sake of the people of Yahweh as it is one that gives redress, relief, comfort, and restoration after a long period of suffering. Again, the dictionary definition is relevant: "REDRESS may refer either to the act of setting right an unjust situation (as by some power), or to satisfaction sought or gained for a wrong suffered," and also definition 6: "to remedy or relieve (suffering, want, etc.)." Perhaps the most striking illustration of the last nuance of English "redress" is found in the highly lyric poem of Isaiah,

77. Num. 25.
78. Judg. 6:1–6.

chapter 35, where in verse 4 the usual translation, "Behold, your God will come with vengeance, with the recompense of God. He will come and save you," is highly jarring and completely inaccurate. It is a description of the expected deliverance from a long-continued situation of want and misery by the "setting right an unjust situation" or by remedying or relieving suffering. There is not a hint in the poem of satisfaction in the form of seeing opponents or oppressors punished or exterminated.

An even broader new synthesis of thought centering around the concept of the executive authority of Yahweh is found in chapter 59 of Isaiah. Beginning with a vivid description of the crookedness and lack of integrity within—and the consequence in blindness, misery, injustice and violence— verse 15b describes Yahweh as acting in "self-help": the phrase is virtually the same as that used in the conversation between David and Abigail (1 Sam. 25:31, 34), and it describes the armor of the divine king preparing for war. The garment of vindication is poetically in parallel to the mantle of "wrath" and the exercise of that imperium is affirmed over all adversaries from the sunrise to the sunset.

Similarly, in Isaiah 61:1–4, the "day of vengeance" has nothing to do with violent punitive actions against an enemy. The imperium rather is the ground for the events most needed by those who are in the greatest misery: the poor, the brokenhearted, the captives, and the grief-stricken.

In all of these usages from Jeremiah to 2 Isaiah one receives the very strong impression that the new emphasis upon the Imperium of God led to the conclusion that all political sovereignty stood under divine condemnation as an enormity that competed with God Himself. It is true that by 540 B.C., the approximate date of 2 Isaiah, there was left not a single important political state of the old era when the dust of world wars had settled. It is not surprising that the lyric prophecy anticipated the dawning of a new era in which Yahweh would "make your magistrates peace, and your taskmasters righteousness" (Isa. 60:17).

What happened eventually is briefly but brilliantly illuminated in the last occurrence of the root in Old Testament times, Esther 8:13. In this completely "secular" book without a single reference to God, the outcome of a complex conflict in power politics is an imperial edict according to which, "The Jews were to be prepared from this day to deliver themselves from their enemies." Not surprisingly, the result was great jubilation among the Jews, and "Many from the peoples of the land made themselves Jews...."[79] What this royal edict actually meant was the granting of political sovereignty to the Jews wherever they might reside, and a freedom from the regular obligations to the Persian Empire. It is not surprising that

79. Esther 8:17.

the result was a bloodbath, but, nevertheless, "on the spoil they laid not a hand."[80] The difficulties in the Book of Esther are prodigious, but it is most tempting to see (however romanticized in the present form of the story) some act early in the Persian Empire by which the Jews were constituted as some sort of state within a state, not on a territorial basis, but on a religious-ethnic one. One thinks, for example, of the *politeuma* of the Hellenistic period, particularly since the term appears in Philippians 3:20: "But our commonwealth is in heaven. . . ."[81]

The Kingdom of God cannot be identified with a political monopoly of force, and the rule of God cannot be identified with formal acts of persons which are induced by hope of gain or fear of punishment.

Similarly, the community is not one which is brought together by common interests of power or profit, nor is it held together by threats of society or rewards in the form of social position, prestige, or power. Nor does God delegate to any political institution sovereignty over persons,[82] and all law can do is merely define (and therefore in part create) evil.[83]

It is depressing to realize how utterly remote the whole concept of the Kingdom of God is from public attention in modern civilization, since, as a matter of fact, it is precisely such voluntary participation of persons in what they believe to be good, right, and necessary that makes it possible for any social order to exist. Unfortunately, it seems certain that the revaluation of such a basis for society takes place only as a last resort—after the blind faith in force has had its usual consequences, so bitterly described and in large part so tragically fulfilled in the biblical predictions of the punitive vindication of God.

Perhaps it was the great disillusionment of the common people with the Maccabean Wars and the corruption of the Maccabean state which more than anything else really prepared the way for the Kingdom of God. At any rate, there can be no doubt that the rediscovery of the relationship between the Old Testament message and human history—and the reinterpretation of that relationship—can be seen in a number of New Testament passages which have to do with the Imperium of God. There are fourteen occurrences of the

80. Esther 9:15.

81. For a discussion of the meaning of *politeuma*, see A. H. M. Jones, *The Cities of the Eastern Roman Provinces* (Oxford, 1937), p. 176.

82. Among other things, this is a primary intent of the narrative in Gen. 1. The New Testament position, to be sure, reaffirms the divine ordinance of the "powers that be," but for a restricted specific purpose, namely, to exercise force (the sword) as the agent of the punitive vindication of God's wrath. This position of course derives from such passages as Isa. 10:5: "Ho Assyria, rod of my anger . . ." It is of course only a radical paganization which can make of such passages the basis for a doctrine of the "divine right of kings."

83. Rom. 7.

New Testament Greek word *ekdikeō*, which was the normal and usual translation of Hebrew *nāqam*, though there are of course many variations.

The only relevant passages in the Gospels[84] occur in Luke: the parable of the importunate widow in 18:1–8 sounds like a translation of Old Testament usage, when she says to the cynical judge: "Deliver me from my adversary . . . ," and the judge finally acts, not out of principle, but merely to be rid of her persistence. The argument *a minori ad maius* is that God will be infinitely more concerned for his elect who appeal (=Hebrew *ṣāᶜaq* 'cry out') "day and night." The passage is strongly reminiscent of Jeremiah 15:15, particularly the perplexing, "Will he delay long over them?" (RSV). This is a reference to Jeremiah's "In thy forbearance take me not away . . . ," which means that the long sufferance of God must result in the death of those who suffer reproach for God's sake.

The role of Jeremiah is also strongly suggested in the other reference to *ekdikēsis*, Luke 21:22. The "signs of the times," including the persecution of the spokesmen for God, the wars of nations, earthquakes, famines and pestilences, terrors and great signs from heaven—all are all too familiar from Old Testament prophecies, as are also the betrayal by close associates and the hostility suffered by those who speak for God—in fact, it almost sounds like a summary of the historical contexts in the Book of Jeremiah in which the old concept of the vindication of Yahweh suddenly took on new meaning. These all lead up, then, to the final climax: the siege of Jerusalem (verse 20), the fall, destruction, and the coming of the Son of Man. It is difficult to avoid the conclusion that the appeal "And let those who are inside the city depart . . ." is taken directly from Jeremiah 51:45: "Go out of the midst of her, my people! Let every man save his life from the fierce anger of the Lord." The fact that the passage is a description of the fall of Babylon makes it even more poignant that the wrath/punitive vindication of Yahweh falls upon *all* of His adversaries—whether Jerusalem (cf. Ezek. 24:2–14) or Babylon. In precise Old Testament fashion, those days are identified as the days of "vengeance." It is curious that Luke alone preserves these strong reminiscences of Jeremiah, where Matthew and Mark have cryptic references to the "abomination of desolation" and the hint ("let the reader understand") that plain language is avoided for sufficient reasons (Matt. 24:15; Mark 13:14).

---

84. However, an omission in the passage of Luke 4:19 is of great importance, and without parallel in the other Gospels. As J. Jeremias has pointed out (*Jesus' Promise to the Nations* [Naperville, Ill., 1958], pp. 41–46), Jesus ended the reading from Isa. 61 in the middle of a verse, omitting "the day of vengeance of our God." Whether or not Jeremias's interpretation of the violent reaction is correct, there is no doubt in the light of the present study that the "day of vengeance" had nothing to do with an expected divine wrath upon the Gentiles, since it here can mean only the *deliverance* of the Lord = "year of the Lord's favor."

The plain fact, from this evidence, is that Jesus did predict the destruction of Jerusalem in no uncertain terms, following the Old Testament prophecies and using to a considerable extent their language and even their grounds. Furthermore, there was, as the "little apocalypse" shows, a great expectation of the death throes of empires and kingdoms, to be followed by the coming of the Son of Man in glory: the historical record of Jeremiah's time, and the great expectations of Deutero-Isaiah (plus Daniel) must have furnished all too firm grounds for such conviction by anyone who took the Old Testament seriously (rather than for proof texts), as the early Christians and Jesus Himself most certainly did. Certainly there was expectation that God would vindicate His elect, but the expectations were dashed to the ground. All that was fulfilled was the destruction of Jerusalem and the temple; Rome and Parthia, which divided the world between them, remained unscathed—just as Babylon was not subjected to the great holocaust which some of the prophets expected.

The unexpected happened; whether or not it was predicted remains perhaps as obscure as the event itself. The reappearance of Jesus to those who valued Him placed a whole new light on the matter, and shortly afterward Stephen describes the vindicative act of Moses in Egypt as the "deliverance [of God] by his hand."[85] The consistent rejection of those sent by God in order to "rejoice in the works of their [own] hands" (Acts 7:41) finds its culmination in Stephen's speech—and in his death filled with the vision of what IS—but not yet.

With Revelation 6:10, we find again the age-old echo of Deuteronomy 32:43: "O Sovereign Lord, holy and true, how long before thou wilt judge and avenge our blood on those who dwell upon the earth?" (RSV). And the answer is that there must be many more before the muster roll is complete, but the celebration in heaven is described in 19:2: "And he has avenged on her the blood of his servants," again a reminiscence of the bitter condemnation of Jezebel (2 Kings 9:7).

The rest of the New Testament passages are largely readaptations of Old Testament sources, whether the warning against self-help on the ground that this is the prerogative of God alone (Rom. 12:19; Heb. 10:30), or the affirmation of God's punitive vindication against His enemies (2 Thess. 1:8; 1 Thess. 4:6), or, finally, the affirmation that even heathen powers are agents of God in punitive vindication (Rom. 13:4; 1 Peter 2:14)—but this is nothing new, in view of the "identification" of Assyria, Babylonia, and the Medes as similar agents of Yahweh. The fact is that the New Testament does not, except by a radical misunderstanding of the whole biblical tradition, justify the idea that a monopoly of force is, in the ancient pagan fashion, a delegation

85. Acts 7:25.

of divine power and authority. They are merely agents, and will be judged as such, for certain purposes. To look to any political power for "salvation," well-being, ultimate security, or the highest good of man or human civilization, which characterized the first half of human history, is the ultimate challenge to God. It is the ultimate idolatry which has been tried and found wanting for so long that only the most blind stupidity and the most determined ignorance can explain the constant determination of men to dehumanize themselves and reduce each other to ashes. The rejection of God is the rejection of love and, ultimately, of life.

IV

THE INCIDENT

AT BETH BAAL PEOR

The rather sordid incident narrated in characteristically laconic style in chapter 25 of the Book of Numbers furnishes an excellent example of the way the new Yahwistic religious covenant structure functioned in history. At the same time, this body of data will be treated here as an illustration of the contrast between the old literary, form-critical method, and what the writer conceives to be a much more relevant method based upon historical, comparative, and functional procedures. As I have argued elsewhere,[1] literary criticism can do little more than reconstruct a history of the written narrative; to proceed from written narrative to real history, it is necessary to appeal to factual data usually beyond the written traditions of the Bible itself.

The biblical narratives rarely if ever give any description of the real-life context of the event described in words. It is frequently the missing context—of cultural patterns of thought, of relationships between spheres of experience as ancient man saw them—that makes much of biblical narrative unbelievable, grotesque, and even absurd. We have no a priori right to assume that people of the Early Iron Age thought and acted just like rabbis of the Roman Empire, theologians of the Middle Ages or the nineteenth century, or professors in modern departments of religion and theology. As a result, every bit of data in an ancient narrative must be viewed against all the evidence we can muster for the purpose of finding the range of ideas and associations which that item of historical fact had in ancient life. It is for this reason that the comparative method is not only legitimate, it is essential. After all, every translation of the Bible into modern English is based upon an unconscious presupposition that the range of meaning of an ancient word or act compares often enough with the range of meaning of modern words to make translation possible. The extent to which we deceive ourselves in this unconscious presupposition is not known until we become aware of hitherto unknown

1. *BANE*, pp. 29–30.

contrasts in meaning between the two vocabularies. Biblical fundamentalism, whether Jewish or Christian, cannot learn from the past because in so many respects the defense of presently accepted ideas about religion is thought to be the only purpose of biblical narrative. It must, therefore, support ideas of comparatively recent origin—ones that usually have nothing to do with the original meaning or intention of biblical narrative because the context is so radically different.

On the other hand, often matters of life and death are involved in the context of biblical narrative, and it takes an experience of similar nature, or good historical imagination to understand the impact of such a situation upon human thought and motivation. Such a situation is involved in the episode at Beth Baal Peor, and certainly at the end both sides were caught up in a predicament of utmost seriousness. It must be admitted that the narrative as we now have it includes materials which come from a period much later than the early twelfth century B.C., and typically the later restatement of the narrative is dominated by interests quite different from those involved in the original event. It is axiomatic that religious *use* of history to achieve some goal or prove some point *relevant* to the interpreter's contemporary situation has almost always taken precedence over the concern to understand historically the original event—if indeed the latter concern was present at all. One need only think of the widespread demand for "relevance" in theological circles today.

In other words, the historical fact of an extremely critical situation has been subordinated by later use to the concern for demonstrating the legitimacy of a priestly line on the one hand, and the dangers of participating in foreign cults on the other. There can be no doubt that the episode became a paradigmatic example in perpetuity, for it is referred to in Joshua 22:17, "Is the iniquity of [Baal] Peor too little for us, from which we have not purified ourselves to this day?" Hosea 9:10 gives a very interesting variant which has no basis in our narrative, and can only come from ancient tradition of another source, "They came to Baal Peor and were made Nazirs to Bošet." Finally, Psalms 106:28 gives still further interpretation of the episode, "And they were yoked to Baal Peor, and they ate sacrifices of the dead." The Deuteronomic tradition, on the other hand, seems to know nothing beyond our existing narrative, or at least adds nothing significant.

The historical context from which the event must be interpreted is the outbreak of an epidemic disease, most probably the black (bubonic) plague. It is necessary to go back to the eighteenth century to find adequate description of the reaction on the part of society to such an outbreak (Daniel Defoe, *A Journal of the Plague Year*). We have no reason to believe that ancient man reacted any less violently to epidemic. Indeed, such epidemics may have been one contributing factor in the large-scale migrations into fringe areas which took place at the transition from the Late Bronze to the Early Iron Age. Virtually every source we have from the Late Bronze Age indicates epidemics;[2] of course, the best known is the Plague Prayer of Mursilis and the related literature, and it is of considerable importance that the reason for outbreak of plague was finally identified as the Hittite king's breach of covenant with Egypt.[3] The conclusion was correct, but medically for the wrong reason. In all probability, prisoners of war returned from the ill-advised campaign against Egyptian-held territory in Syria brought the plague with them. There are also references to the plague in Ugarit, in the Amarna letters from Palestine and Cyprus, whose king requests from the pharaoh one of the "bird-watchers" (šāʾili našri), most probably for the purpose of divining the cause of the plague so that appropriate propitiatory action could be carried out.

Some pestilential disease was endemic to the entire region during the period with which we are dealing (ca. 1400–1000 B.C.); the biblical narrative at Joshua 22:17 above indicates that at some subsequent period the plague was still present in Israel. The narrative of the Philistine return of the ark from Ashdod to Beth-Shemesh strongly suggests bubonic plague and a knowledge of the association between rats or mice[4] and the pustules which characterize the Black Death. It is intriguing that there are also very close similarities between the Philistine ritual action in returning the ark, and liturgies for the removal of plague found in Hittite sources.[5]

Our sources seem to show the following response on the part of officialdom to the outbreak of epidemic disease: a consultation of the gods through various kinds of specialized divination to ascertain the cause of the outbreak usually identified as a moral or ethical delict, a confession of sin, and appropriate action as a propitiatory ritual in order to remove the cause of the wrath.[6] Within the framework of what is presently known for that period, such a context for the Baal Peor episode is entirely satisfactory as a starting point for its interpretation. From the fact that the plague did break out in the course of the religious ritual we can be sure that it was already present among the "Moabites." This is also reinforced by the traditions of chapter 31 of Numbers, which describe in detail the extreme care taken to

2. E.g., *EA* 35:37 Cyprus, and 96:6–18 Ṣumur; F. Thureau-Dangin, "Nouvelles lettres d'El-Amarna," *RA*, 19:91–108, AO 7093, ll. 45–50, where Rib-Addi denies reports of plague in the country and labels them as attempts to dissuade the king from sending troops; *EA*, 244:32 Megiddo; and cf. below.

3. Translation in *ANET*, pp. 394–96.

4. Or even voles.

5. On the ritual driving of an animal to the enemy country whose god has caused the plague, see *ANET*, p. 347. It is specifically Arzawan = Luwian.

6. For a remarkably similar procedure still operative in society, see Middleton, *Lugbara Religion*.

avoid spread of the plague after the battle against the Midianites. The entire narrative complex is historically authentic in virtually every detail preserved, regardless of literary "source," but it is quite clear that the *use* of the narrative is dedicated to purposes that stem from later contexts and concerns. On the other hand, we can see certain side effects of the episode that had lasting historical implications, and it seems probable that the event itself served very powerfully to reinforce the newly established covenant ideology and community that began at Sinai and spread over much of Transjordan and Palestine in a generation.

The event may be reconstructed along the following lines: various tribes or bands which had fairly recently migrated south to Transjordan from the north or northeast either formed or joined an already existing confederation that was called Midian.[7] It was subjected perhaps willingly to the suzerainty of Sihon, "king of the Amorites,"[8] and after his defeat by the newly formed Israelite confederation (the covenant at Shittim, preserved in traditions as the "second giving of the law"),[9] the Midianites formed part of the population ruled by the king of Moab[10] (there was no such thing as a "Moabite" except in the sense that anyone recognizing the legitimacy of the Moabite king or residing in territory ruled by him might be so designated by outsiders). It is entirely possible that the recent immigrants brought the plague with them from north Syria or southern Anatolia, and it is even more probable that they brought with them the camel and rapidly developed a nomadic type of culture.[11] Quite close relationships had existed between Israel and the Midianites since the days of Moses' sojourn in the Sinai desert.[12]

With them from the North (or with other migrants) came the cult of Baal Peor, just as Baal Zaphon was localized along the chain of lakes between Egypt and Sinai. The mountain and town cannot yet be identified for certain, but they cannot have been very far from Jebel Siaghah or Mount Nebo

7. See chap. VI.
8. Cf. Josh. 13:21. The same list of names occurs in another context in Num. 31:8, where the title "king" is given them. In view of the fact that the root *NSK* (*nasîk* of Josh. 13:21) has to do primarily with libations, which we know to be a Late Bronze Age custom by which a "great king" conferred local authority upon a vassal king (*EA* 51:4–9), it seems virtually certain that the authorities in question were local chieftains (= "king") whose social power was dependent upon a superior one. We know also from the Mari documents of the eighteenth century B.C. that heads of small north Syrian societies claimed the title "king."
9. Also specifically identified as a covenant in Deut. 28:69 (29:1 Eng.).
10. Compare Num. 22:4, 7 and 22:14, 21.
11. Judg. 6:1–6.
12. Exod. 2:16–22; 18:1–24. It is difficult to see how such traditions could have been motivated by anything other than historical fact.

and the Wadi Uyun Musa. The origin of the Lord of Peor seems now clear. There is no satisfactory Semitic etymology, in spite of the fact that a Ugaritic root with the same consonants occurs with the meaning "to call by name,"[13] or something similar. The *name* occurs in the Karatepe bilingual inscription with precisely the Hebrew spelling in the Phoenician version, and as *Paḫura* in the hieroglyphic Luwian version.[14] The latter is of course the usual Hittite word for "fire,"[15] and is a form of the root that underlies Greek *pyr* and eventually English "fire." The god in question is simply "lord of fire," which also provides the place name. A town is attested near the pass which leads from Alexandretta to Antakya that still has the name Bağra and that in the Middle Ages was called Pagrai—similar to the Septuagint spelling of Baal Peor.[16] From the Semitic point of view, a "god of fire" is inevitably closely related to the plague,[17] but one thinks also of Apollo, and the mysterious references to the "burnings" caused by Yahweh in the camp[18] may well be late and misunderstood disguised references to minor epidemics —we just do not know.

13. *UT* Glossary, p. 469. Cf. Hebrew p^cr 'open the mouth wide,' all late uses except Isa. 5:14, where it is clearly associated with the actions of a chthonic deity, here Sheʾol.

14. R. Marcus and I. J. Gelb, "The Phoenician Stele Inscription from Cilicia," *JNES* 8 (1949): 116–20; C. Gordon, "Azitawadd's Phoenician Inscription," *ibid.*, pp. 108–15. Luwian *Pa-ha + r-wa-ná-i*, VILLE according to *HH*, p. 177.

15. *paḫḫuwar, paḫḫur* (*HE*, 2:95). Luwian *paḫur* (*DLL*, p. 77).

16. LXX *Phegor* presupposes of course the strong final accent of Palestinian West Semitic. *Pagrai* is the sequel to Assyrian *Paḫri* (Marcus and Gelb, "Phoenician Stele Inscription," p. 118, and n. 8). For the Crusaders' *Pagrai*, see R. Dussaud, *Topographie Historique de la Syrie antique et médiévale* (Paris, 1927), pp. 433–35. It was already important in the Seleucid period, and would make an excellent candidate for the southeast boundary fortress of Azituwadda of the Karatepe inscription.

17. Cf. Semitic Rešef, god of fire and plague. The Karatepe *ršp ṣprm* (see Marcus and Gelb, "Phoenician Stele Inscriptions," and Gordon, "Azitawadd's Phoenician Inscription"), which has been interpreted as Rešef of the Wings, or Rešef of the Bucks, could well be merely Rešef of the town called *ṣprm*. Whichever interpretation is correct, Balaq in the Baal Peor story is called 'son of ṣippor,' which establishes still another strikingly close connection between the Luwian territory of Cilicia and the names of early Transjordan. Cf. also Ugaritic *ṣpr, ṣprn, ẓbr*, and *ṣu-pa-ra-nu* (*PTU*, pp. 190, 183). Other than as the name of the presumably North Syrian Moabite king, the name occurs in biblical Hebrew only in the feminine form; as the name of Moses' wife, *ṣipporah*. She was of course the daughter of the "priest of Midian," and therefore also of Anatolian-North Syrian origin ultimately (see chap. VI).

18. Num. 11:1–3, where the narrative is evidently etiological—to explain the place name *Tabᶜera*. The Hebrew root *Bᶜʀ* 'burn' has no convincing Semitic etymology, and may possibly be still another old divine name which has become a common noun or even verb. For similar traditions about the plague god's "burning" opponents, see the story of Coronis (chap. III, n. 66) where the outbreak of plague is *compared* to a forest fire.

When, therefore, the epidemic broke out among the Midianite tribes, an appropriate ritual was instituted in order to remove the "wrath." What that "wrath" was thought to be, we do not know, but the ritual resorted to consisted as usual of sacrifices to the deity whose protection from the plague was desired, and then ritual sexual intercourse with outsiders. We have no direct knowledge of relationship between the latter and apotropaic rites; however, though it cannot yet be proven, there are two lines of evidence that would strongly suggest a connection.

First, we have the evidence of the Greek Anthesteria and the very early ritual evidence from Mycenean Greek texts which has already been pointed out by L. R. Palmer[19] as probable extensions of very ancient Near Eastern customs to Mycenean culture. Since the Anthesteria was dedicated to the appeasement of the spirits of the dead, it seems entirely likely that the Baal Peor ritual was likewise determined by this sort of ideology. As is well known, ritual intercourse was the procedure carried out on the second day of the festivities. The first day, in Mycenean, was called, "the opening of the jars," which exactly translates the name of a festival Yasmaḫ-Adad wished to carry out at the Mari palace, but was instructed not to by his "father" Šamši-Adad.[20]

If ritual intercourse is connected with the necrolatry of Mycenean culture, in the Sumero-Babylonian *Utukkū Limnūti* texts we also have evidence of connections between disease and the unappeased spirits of the dead.[21] The texts had already become a quasicanonical manual for exorcism by the time of the Old Babylonian period (in Sumerian with interlinear Babylonian translation). These traditions, rituals, and underlying patterns of thought which identified illness with the consequences of being "seized" by some malevolent spirit of the dead, exhibit an extraordinary range of distribution, no doubt from prehistoric times, and probably even underlie certain aspects of popular religion in New Testament times.[22] Among those spirits of the

19. L. R. Palmer, *The Interpretation of Mycenean Greek Texts* (Oxford, 1963), pp. 250–56.

20. *ARM*(T), I, 52:9–11. The phrase is obscure (cf. Wolfram von Soden, *Orientalia*, n.s. 21 [1952]: 80; *AHW*, pp. 730–31, 777). The passage certainly has something to do with the royal table, as is shown by ll. 31–35. Cf. also *naptan šarrim* (*AHW*, s.v. *naptanum*) as a cultic term. A similar ritual term has recently been pointed out in a paper at the American Oriental Society in Baltimore in 1970 by R. Carter from Hittite sources: "opening of the jars."

21. Series *utukkū limnūti* (cf. A. Falkenstein, *Haupttypen der sumerischen Beschwörung* [Leipzig, 1931], pp. 38–39), where the series under discussion is classified as the "prophylaktischer Typ."

22. Mark 5:9; Luke 8:33. Cf. especially the association of the demon-possessed (whose name is "Legion") with the residence among the "tombs." The event took place, of course, in Transjordan. Note also the "spirit worship" of the Lugbara (Middleton, *Lugbara Religion*).

dead that returned to haunt the living were, "a girl who did not reach her bloom" and "a boy who remained unyoked."[23] The Babylonian version of the second quotation uses the same verb[24] that is used in Numbers 25:3, and it is beyond question that it has sexual connotations. The underlying idea is the very primitive one that the boy or girl who went to death without having experienced sexual intercourse remained unsatisfied and therefore caused harm to the living. If the necrolatry consists of offerings by the living of those things which cause the dead to be made happy, then the sequence of sacrifice, food and drink, and ritual intercourse would represent the gamut of those things necessary to put the restless spirits at ease.

The terminology used does not make it possible to determine whether or not the "daughters of Moab" are unmarried maidens who are thus ritually sacrificing their virginity.[25] Most often in biblical Hebrew the term "daughters of . . ." followed by a place name seems to designate unmarried females, but this is not predictable.[26] Later traditions indicate that in Babylonia (and Syria) every female had to prostitute herself,[27] but such narratives, coming exclusively from Greek or later sources, lack context even more than biblical narratives, coming from a much later period after radical changes had taken place in traditional societies, and cannot be relied upon to interpret narratives of the thirteenth century B.C. It remains a curious fact, however, that biblical narrative illustrates both ritual intercourse and ritual murder (human sacrifice) only in Moab and Gilead,[28] and it is here that we have the most massive onomastic evidence of migration from Anatolian sources.[29]

23. My thanks are due to Thorkild Jacobsen, who has very kindly helped with the interpretation of this difficult text. Sumerian: *ki sikil šu nu.un.du₇.a* [. . .]*guruš á nu.un.lá.e*[. . .], Babylonian: *ardatu la šuk[lultu. . .] etlu la su-um-[mu-du. . .]* (*CAD*, 16: 247). Cf. Falkenstein, *Haupttypen*, p. 38, n. 2.

24. ṢMD. Noth commented upon the peculiarity of the phrase, and tried to make a *Namensätiologie* out of the narrative (Martin Noth, *Überlieferungsgeschichte des Pentateuch* [Stuttgart, 1948], p. 81); later he seems to have given up the theory (*Numeri*, ATD, p. 171). Roots with the normal meaning "to yoke together" have sexual meanings in virtually all the Semitic and Mediterranean languages. For an attempt at interpretation of this passage, see below.

25. Ritual sacrifice of virginity is widely known. See n. 27.

26. E.g., 2 Sam. 1:20, 24; Gen. 36:2; 6:2; Isa. 3:16. Several passages prove that *bᵉnōt-* followed by a gentilic or place name designates those who are eligible for marriage, but it cannot be proven *always* to have this meaning.

27. Herodotus *Persian Wars* I. 199; cf. Lucian *de dea Syria* 6.

28. The narrative of Eli's sons at Shiloh (1 Sam. 2:22) contains no indication of any particular cultic significance. It seems probable that such customs were reintroduced during the Solomonic temple, as indicated by the periodic reforms of Asa and other kings (1 Kings 15:12; 22:46; 2 Kings 23:7). Cf. Deut. 23:18–19; Hos. 4:13–14. So far as human sacrifice is concerned, the less said the better, until we have texts which describe the meaning of "causing [a son] to pass through the fire." Fire rituals are too common to justify beyond doubt the conclusion that human sacrifice is involved. Further, note that

There are other questions that might be asked but cannot at present be answered. Why must it be a stranger who is invited to participate? Why is it the women whose participation is necessary for the proper performance of the ritual? One suspects that the answers to such questions may derive from psychology more than from history, and there is little point in further speculation. Another question that cannot be answered is whether or not the ritual may simply have been a seasonal one, like the Anthesteria, but there is no reason to believe that the seasonal and the crisis rituals must be mutually incompatible. At any rate, it seems entirely justified to conclude that the feast followed by ritual sex acts was a patterned ritual behavior the meaning and purpose of which was known to all concerned. It could well be the sort of thing that had happened before on various occasions.

There is only one new factor which determined both the interpretation of the event and subsequent action from the point of view of ancient Israel. The ritual constituted breach of covenant, in that it either created or presupposed a relationship to a pagan god. The reference to "yoking" we take to be a designation specifically of the ritual intercourse, performed for or in deference to Baal Peor—it was not the deity to whom they were "yoked," though Hosea perhaps interprets the act in this way ($yinnāz^eru$).[30] Nevertheless, the ritual of healing diseases caused by the unrequited spirits of the dead characteristically takes place by appeal to a high god to intercede in protecting the living from the dead. In addition, there is often a ritual identification of living participants with the god, though this usually involves a priestly exorcist. A characteristic limitation of the comparative method is the fact that though it may well provide us with the broad context of functional meaning, it can seldom enlighten us on specific details, which are usually much too closely tied to the culture that gives them meaning. Such rationalizations of cultic language and acts are usually much more important to the philosopher and theologian than they are to the participants, who are much more concerned with the crisis and the cure than they are with metaphysical and philosophical interpretation of meaning.

The whole event gives us a historical paradigm of the covenant pattern of thought as a basis for the interpretation of the event, and the action which ensues is given rationality only upon such presuppositions. If ancient Israel as a community consisted of those who had bound themselves in typical

---

in 2 Kings 16:3, this ritual is specifically derived from the "nations whom the Lord drove out . . ." and which we can now definitely identify in large part with Anatolian populations. See chap. V. In addition, 2 Kings 17:31 specifically links the "burning" of children to the introduction of the Sepharvites, who seem likely to be from North Syria or southern Anatolia.

29. Chap. VI.
30. Hos. 9:10.

Late Bronze Age fashion to obey certain stipulations at the cost of incurring the curses brought about by the divine power, then the violation of the first command was followed in a direct, undeniable sequence by the imposition of one of the most important of the curses: pestilential disease. Though the point should perhaps not be pressed, it is nevertheless worth observing that the usual incubation period of bubonic plague when infection takes place through insect bites is about seven days.[31] On the other hand, when the bacillus is taken into the lungs from the breath of an infected person, the disease may run its entire course to death within a period as short as a day or even less.[32] The rituals involved would guarantee the maximum probability of such an infection's leading to what is known as pneumonic plague. Particularly to a relatively unsophisticated group,[33] the dramatic and fatal association between violation and death by divine curse would be most impressive. But under the conditions of mass participation in such spread of disease, the guilty and innocent alike would suffer, and the only ways of reacting appropriately would be, first, cessation of the violation, and, second, dissociation of the community from the guilty by imposition of the death penalty. It is not certain that the death penalty would have to have been imposed,[34] but it seems overwhelmingly probable on the ground that the community which refuses to punish the guilty must share in the consequences of the guilt. Since the consequences by universal consensus in the ancient world included death by plague, there would be little point in the community's protection of the guilty.[35]

The reaction to the "wrath of God," which certainly can be identified with the outbreak of the epidemic in the Israelite "camp," whatever that may have been at this time, is not described in satisfactory detail. In fact, the narrative at this point is quite disjointed, as Noth has pointed out.[36] Verse 4 describes Yahweh's instructions to Moses to "take all the heads of the people and *hôqaᶜ* them to Yahweh in the sun, that the wrath of Yahweh may be averted from Israel." But the words of Moses which follow are almost totally

31. I. L. Bennett, Jr., and E. S. Miller, "Plague," in *Principles of Internal Medicine*, ed. T. R. Harrison *et al.*, 5th ed. (New York, 1966), art. 289. It varies from one to twelve days.

32. Death from pneumonic plague occurs in one to five days (*ibid.*).

33. Particularly since they had been sensitized to the possibility in advance in the covenant formulas.

34. The covenant form, of course, provided *only* for divine sanctions, not for any imposed by social organization or force. This is one of the many extremely important distinctions between covenant and law, most of which have not been seen in modern scholarly literature.

35. There would be as little point in punishment: the infected would normally die anyway. All that is necessary is disassociation, but in case of epidemic this is impossible.

36. *Numeri.*

unrelated to the instructions of Yahweh. Furthermore, there is no statement of compliance; rather a narrative of an individual act in an individual case. The entire exposition is so uncharacteristic of late narrative form (the omission of the statement of compliance would be virtually impossible by the time of the Priestly Code)[37] that it can hardly be explained except by the hypothesis that we have several fragments of tradition, all early but not mutually consistent. This is itself a fairly good indication of authenticity; the disjunctions were permitted to stand because later tradition did not know how to harmonize them. In Yahweh's instructions it seems that all the "heads of the people" were to be executed for the sins of the guilty. Moses' instructions, on the other hand, appeal to the "judges" to execute only the guilty participants in the cult. There are too many possibilities for interpretation of the inconsistency, and the information given in *both* verses 4 and 5 is highly relevant to the twelfth-century situation. This entire narrative has few if any parallels in the biblical tradition in the following respects: (1) The instructions of Yahweh to Moses presuppose the idea of representative punishment (i.e., vicarious punishment of the innocent for the crimes of the guilty, presumably because the innocent have positions of social power and have not restrained those under them). (2) The instructions of Moses to the "judges" presuppose exactly the opposite: the obligation of those in power to impose punishment directly upon the guilty, and Moses' words have virtually nothing in common with the instructions of God. (3) Neither of the foregoing traditions is connected with a statement of compliance, which was an extremely binding literary form in the late stages of tradition formation.[38]

The resolution of the dilemma presented by the preserved tradition is perhaps to be found in the reconstruction of historical probability. The "Mosaic law" in its later stages is formulated after the model of royal law, and thoroughly presupposes that Moses acted exactly like the kings from David and Solomon on—which is a priori the least likely historical probability. Paradoxically enough, the law which the king is supposed to promulgate, according to the conservative religious party tradition, is one which originated before the Monarchy—a tradition which I believe to be most emphatically correct. It follows that a radical contrast must be made between the substance of the law and the enforcement procedures. The former stems from Moses and the Mosaic tradition, the latter comes from the Monarchy; and they are mutually incompatible. But neither can cope with the authentic history of the incident at Baal Peor. In this episode which took place before the nascent religious community was a generation old, the people had to make a serious choice of action on the foundation of a new system of religious

37. Noth also points out that the narrative in 6–18 cannot be P.
38. See the forthcoming Numbers in *The Anchor Bible*.

convictions and criteria for judgment and there was no adequate precedent.

There was no accumulated body of empirical evidence which could prove in a law court (if such existed in the modern sense of the term) that the Midianite techniques for coping with the Black Death were efficacious. There existed no body of tradition or evidence that could prove that pagan traditional techniques for averting the plague were wrong. There was only the fact that those who constituted "Israel" had sworn allegiance to Yahweh and to the stipulations which were inseparable from that allegiance: in fact, there was no meaning to the allegiance other than obedience to the stipulations. (It was not until centuries later that allegiance to Yahweh was defined liturgically.) It is not too surprising, therefore, that some liberal-minded conservative decided that the tried and true methods were most probably correct, and brought a willing Midianite maiden to his peer group in order to appease the unrequited spirits of the untimely deceased. There have been a number of dramatic confrontations of theological interpretations in biblical history. This I believe to be the first historically accessible one; the second was at Mount Carmel; the third between Jeremiah and the determined paganists after the destruction of Jerusalem; and the last really important one between Jesus and the priesthood prior to his crucifixion (with a relatively minor repercussion later in the confrontation between Peter and Paul at Antioch). Always the issue is the same: is the community to trust a relatively new complex of ethical convictions or the age-old accumulation of ritual pseudoscientific techniques? The political system, in order to be just, *must* follow the latter, for it cannot impose as a criterion for the application of its monopoly of force an ethical judgment which was not generally accepted and recognized at the time legal obligations were willingly assumed. Law *must* presuppose an *existing* ethical value system illustrated in customary behavior. The fact that that system may be both inadequate and intolerable to many members of society is irrelevant to law, but far from irrelevant to those members of society. It is much easier to reject the existing legal system and its values than it is to put a superior one in its place. That is what ancient Israel did—and in the process they had to reject existing legal structures. The early New Testament community did the same without using violence to secure its immunity from the jurisdiction of existing political organizations: it conquered but was conquered by them in the new synthesis which led the way to the Middle Ages.

It is quite possible that the Simeonite Zimri was very sincerely convinced that his ritual conduct with the Midianite princess was the best way to secure the welfare of his group.[39] It is absolutely certain, if present theories of the covenant structure are correct, that he was assuming that concerns other than

39. The narrative clearly implies that the plague was actually epidemic in the Israelite camp.

a sworn promise of obedience should take precedence when matters of life and death are involved. It is easy now to condemn this Simeonite, when we have far more technical knowledge of the means of infection and course of the black plague bacillus. The fact remains that he was acting in conscious rebellion against that which he had promised to obey—if he were not, he would not have been identified with the infant community of ancient Israel.

Curiously enough, though the value system of the ancient Israelite community was the diametric opposite of everything that would have been regarded as "scientific" at that time, the actions of the Simeonite were such as would guarantee the maximum spread of the epidemic. Convinced of the validity of "age-old" remedies, he acted in a way we know most efficiently to increase the malady. The consequences to society were precisely those predicted by the covenant formula, and the case became a parade example even in New Testament times.[40]

In such circumstances, what should a community do? It is curious that we have three different reactions to the crime: the instructions of Yahweh evidently represent the reaction of some religious establishment which we can neither identify nor date.[41] The instructions of Moses are quite different,[42] but it is the extremely modern Phineas who takes action.[43] We do not, in fact, have any historically plausible narrative which would indicate that the covenant community before this time conferred upon anyone the authority to put persons to death for violation of the covenant. Phineas represents the transition from covenant to law: from the enforcement by Yahweh to the enforcement by human action. In this case there is no social sanction for his behavior, though it is evident that subsequent tradition represented in this part of the narrative must have approved.[44] He murdered not only the Simeonite, but also the Midianite princes in *flagrante delicto*.

40. I Cor. 10:8, with a curious reduction of the fallen by a thousand.

41. We have no good parallel for the execution of social authorities for the sins of their subordinates—the assassination of various kings who were blamed for social ills seems an improbable source for the tradition.

42. It is tempting to relate the two: the "judges" must execute the punishment if they are to avoid the consequences themselves. In other words, the concern for the enforcement of law and punishment of crime must take precedence over the normal concerns of tribal societies to protect their members at all costs. This is exactly the issue in the narrative of the rape of the Levite's concubine in Judg. 19–20.

43. Such individual action is illustrated in the story of Ehud and Eglon, King of Moab, in Judg. 3:15–23 (cf. Jael and Sisera, Judg. 4:17–21). The assassination of kings, especially in Israel, may well be a continuation of the same very archaic pattern of thought and action that correlated to the early lack of highly developed social instruments for the peaceful and orderly solution of conflicts, especially beyond the small primary community.

44. This is merely one of three narratives which derives the authority of the Aaronid priesthood from the Mosaic period. Cf. Num. 17 and Exod. 7.

One strongly suspects that this narrative, though it is usually assigned to the P source, is actually correct. It is the words of Yahweh and of Moses that in their present form have all the earmarks of some ancient scholarly lamp. The transference of enforcement powers from Yahweh to society took place at Baal Peor. This represented the transference from covenant to law, for the society not only condoned (evidently), it also accepted the historical consequences of the individual, unilateral act of Phineas, and, characteristically, made the episode a binding precedent![45] It is hardly possible to conceive of a more heinous crime than the murder of a princess in a temple (or sacred precinct of some sort) engaged in a sacred ritual which was supposed to remove a most serious divine scourge threatening the entire community. From the point of view of the Midianites, in spite of the earlier very friendly relationships and even marital ties,[46] such an act could only mean war. The sequel is found in chapter 31 of Numbers, also traditionally assigned to the Priestly source, which characteristically misunderstood the historical issue entirely, obsessed as they were at that late date with ritual purity under historical circumstances of a radically different nature.[47]

The words of Yahweh to Moses (25:4) are extremely suggestive. The similarity of wording to 2 Samuel 21:6[48] strongly suggests that we are dealing here with a very old formalized idiomatic expression, which in turn correlates with a highly formalized custom. Everything turns on the meaning of the term $hôqa^c$. It is not easy to be certain of its meaning, for the term is used only in three narratives in the entire Hebrew Bible:[49] the two already mentioned, and the narrative of Saul's death in 1 Samuel 31:10, where we should read $tôqî^c\bar{u}$[50] instead of $tāq^{ec}\bar{u}$, with no change in the consonantal text involved. The meaning of a term is what the uses have in

45. Zeal to commit murder in defence of "religious" ideals did not end with Phineas, but it has rather rarely been made the basis of institutionalized authority. The interpretation hardly represents the norm of Old Testament religious thought, and would be most appropriate in the bitter struggles associated with the reign of Ahab and Jezebel.

46. I.e., Moses and Zipporah, daughter of the priest of Midian.

47. See the treatment in chap. III—the "vengeance" of Yahweh.

48. The treatment of Saul's sons demanded by the Gibeonites so described was, of course, the consequence of a divine curse upon Saul for mass murder and a subsequent famine.

49. Somewhat similar procedures are associated with the root TLH (2 Sam. 4:12), and the famous prescription in Deut. 21:22–23. The narratives in Josh. 8:29 and 10:26 use this root to describe the execution of the kings of Ai and the coalition of five kings defeated at the battle of Aijalon. Cf. also the Egyptian hanging of six princes of Tḫsy (Taḫši) on the city wall of Thebes (A. Gardiner, *Ancient Egyptian Onomastica* [Oxford, 1947], 1:151*.

50. Assuming that it is an archaic 3pl impf. form with preformative *t-* instead of *y-*. Cf. *UT*, p. 75.

common so far as we can determine. In the two cases in the Books of Samuel, it is very explicit that exposure of the corpse until final disintegration is intended. The death of Saul proves that the meaning of the term need not include the imposition of the death penalty, for it was a procedure carried out upon the corpses, not upon living persons.

A second similarity which applies to all three cases (our narrative, as stated before, provides no hint that the procedure was actually carried out) is the very significant fact that this punishment was an early form of punishment for breach of covenant. The narrative of the Gibeonites is very explicit on this point: although it is a later editorial addition, it is perfectly correct. The covenant of peace between Israel and the Gibeonites was binding, and Saul's misguided attempt to gain political support among the Israelites and Judahites by murdering the Gibeonites must have been interpreted by the latter at least as breach of treaty. It was evidently so interpreted by Yahweh, and famine is of course a classic divine punishment for breach of covenant.[51] Since Saul himself could not be punished, the fate appropriate befell the hapless descendants—which Saul should have thought of before if he were concerned for his children; for the Decalogue formula "visiting the iniquities of the fathers upon the children to the third and fourth generation . . ." certainly is not late. It is quite possible that Saul cared little about old Mosaic tradition, for according to the genealogical lists of 1 Chronicles 8:29–34 and 9:35–40 his ancestry went back to Gibeon, a Hivvite city.[52]

Similarly, the Philistine treatment of the corpses of Saul and his sons is predicated upon the fact that from their point of view Saul had been guilty of breach of (vassal) covenant. It is conceivable that the reference to *hôqaᶜ* in Numbers, chapter 25, may have been an extrapolation into this narrative *because* the offense was regarded as breach of covenant, but if so the interpretation must have been very early since there is little evidence that the old covenant pattern of thought survived in any important social-political form after King David, and no later attestation for the custom is extant in biblical sources.[53]

A third similarity which all three of our narratives share is less clear but in view of one main theme of this series of essays is nevertheless something

51. Deut. 28:23–24, 17–18. Compare also the drought of 1 Kings 17:1, which is certainly to be associated with the murder of the prophets by Jezebel (1 Kings 18:13), the drought as a consequence of Aqhat's murder in the Danᵓil epic at Ugarit, and the sacrifice of Athenian youth to the Minotaur as expiation for murder to end a drought. It seems to be a universal in ancient Mediterranean societies dependent upon rainfall.

52. In view of the proximity of the two towns, Gibeon and Gibeah, there may easily have been local feuds that gave Saul further excuse, particularly since Saul's ancestors had at some time defected from Gibeon.

53. As noted above, in Deuteronomy the custom of exposing corpses until disintegration is prohibited.

to be pointed out. The punishment is meted out to Saul and his sons by Philistines—a political structure dominated by a complex of Aegean-Anatolian peoples.[54] The punishment is demanded by the Gibeonites, whom the Bible terms Hivvites and whom we have equated with the population of Cilicia.[55] The Baal of Peor is Anatolian, and the Midianites are Anatolian or North Syrian in origin[56]—as are also many of the early Israelites at this time. Rizpah, daughter of Aya—a typically Hittite name[57]—the concubine of Saul, is an early prototype of Antigone who similarly is caught between political policies and personal loyalties, but she solves the problem in a typically Near Eastern way which doesn't make such good tragedy—it causes no further deaths.[58]

We cannot leave this theme without pointing out several historical ramifications of extreme importance. First, this custom of execution and exposure of the body until distintegration is referred to only in the three passages under discussion. The verb never occurs later in either this form (*hif ̔îl*) or with this meaning, and all the passages under discussion are assigned to the time of the United Monarchy at the latest. A second point of importance to this discussion is the "Deuteronomic" law of 21:22–23, where the custom of hanging up a corpse is still presupposed (i.e., of a person executed for a crime subject to a death penalty), but burial the same day as execution is stipulated. One suspects that the vast changes taking place during the United Monarchy caused a complete reversal in this respect. At Baal Peor and Gibeon it is precisely *because* the culprit is guilty of ultimate rebellion against Yahweh, i.e., breach of covenant, that he must be executed and exposed; in Deuteronomy it is *because* the culprit is cursed that he must be buried as soon as possible. At any rate, we hear no more of the custom in ancient Israel. With the demise of the old covenant patterns of thought, the appropriate punishment for violation had no longer a foundation of public opinion—and it was an offense to the eye and nostril. Finally, the prophetic protest would strongly suggest that after all breach of covenant was not regarded as nearly so serious a crime as breach of proper liturgical ritual[59]—and we find the same sort of thing in Babylonian texts where "sin" is frequently coterminous with liturgical neglect.[60]

54. See H. Hencken, *Tarquinia, Villanovans and Early Etruscans* (Cambridge, Mass., 1968), p. 146, for the evidence that Philistines and Etruscans are closely related culturally.
55. See chap. VI.
56. See chap. VI.
57. It is one of any number of names that *can* be interpreted either as Anatolian or Semitic.
58. Sophocles *Antigone*.
59. Isa. 1:12–17.
60. Cf. the ritual for the New Year's festival, *ANET*, p. 334, where the king's confession is heavily liturgical, though not exclusively so.

The prophets, however, evidently continue the old-fashioned ways of thinking. When they so often proclaim that the population will become dung upon the earth, they are saying in effect that though the law no longer enforces this punishment for radical breach of covenant, nevertheless Yahweh Himself will, through what we so blithely call the "historical process." In this way, the distinction between law and covenant is still very much alive in the prophets, even though the law of the state had become dominant—but when the state is gone, nothing is left but covenant.

The story does not end here, however. The data we have lead to the conclusion that the tradition of exposing the corpse and denying it burial is closely associated with Anatolian-Aegean populations at the time under consideration, and typically it is associated with the concept of punishment for the most serious types of crime against society. The Assyrian Empire illustrates the continuation of this kind of punishment: the custom of impaling rebels against Assur. It would take another monograph to demonstrate this thesis, but that this association exists is beyond question.[61] Whether or not impaling was imposed *only* in those situations in which the culprit had flagrantly violated his oath of obedience to Assur and the latter's viceregent, the Assyrian king, is another problem.

Some 800 years after the Assyrian policy of calculated cruelty, we again find hanging as a punishment for crimes against the state under the Roman Empire—crucifixion.[62] It may be nothing more than a curious coincidence, but it is nevertheless true that the episode at Baal Peor in its present form narrates the command of Yahweh to *hôqaᶜ* the "heads of the people to Yahweh" because of the flagrant violation of the people themselves. The crucifixion of Christ is placed into this context very consciously by Paul, and by other New Testament writers as well.[63] Using millennia-old patterns of thought, Israelite as well as pagan, the body of Christ was looked to as that which removes the wrath of God (or the gods, from the pagan point of view); at the same time, since the innocent is put to death for the guilty, those who have benefited are placed under the most powerful obligation to avoid future breach of covenant.

The reciprocal relationship between the individual and the community is most complex and most changeable, depending upon the vagaries of historical event. But two principles are most evident in the biblical and extra-biblical materials: (1) He who protects the guilty is himself subject to the same penalty as the guilty. (2) Until human beings no longer react with

---

61. *ARAB* I: 202, year 28: "These (rebels) I impaled on stakes." There was such a range of variety in the treatment of enemies that the association should not be emphasized.

62. Especially for slaves and rebels.

63. Gal. 3:13.

anger and violence against murder, theft, adultery, and fraud, the acts of individuals are certain to have important and serious effects upon the total human community. Phineas saw both principles, but it took much longer to see that there is a reciprocal to both of these principles: (1) He who does good has an attributive effect upon a wide spectrum of society. (2) It is not so important to bring retribution upon those who violate common decency as it is to create those conditions that will make murder, theft, adultery, and fraud so unattractive to individuals that they will become nonexistent. This *is* the Kingdom of God. It *was* the great vision of early Israel until it became overwhelmingly enamored of the power, the kingdom, and the glory, and the wealth of Solomon, the wisest of fools.[64]

64. Jer. 9:23.

V

THE ʿAPIRU MOVEMENTS

IN THE LATE BRONZE AGE

Two interrelated problems of the Amarna letters have received lively discussion in past generations. One is the meaning and derivation of the term ʿApiru/SA.GAZ,[1] and the other is the question of the possible connection between the activities of the Amarna ʿApiru and the biblical narratives of the "conquest" of Canaan by Israel.[2] In my opinion, it is inconceivable that the Amarna letters should ever have been used as materials for the reconstruction of Israelite history had it not been for the identification of the Amarna ʿApiru with the biblical ʿIvrî. Consequently, for a long time there was a tendency to regard ʿApiru as a gentilic term, and, in the reverse direction, a tendency to reconstruct biblical history on a foundation far different from that presented in biblical traditions. Furthermore, this connection with biblical traditions was based on a misunderstanding of the Amarna period: an assumption that these letters present a picture of nomadic invaders attacking and storming the Canaanite cities of the Egyptian Empire.

It is now agreed by nearly all scholars that the term ʿApiru originally had no ethnic significance, but rather designated a social or political status.[3] It remains merely to apply this insight and to reexamine the events of the Amarna age in the light of this fact. If the ʿApiru were not nomadic hordes of barbarian invaders, who were they?

The ideogram SA.GAZ[4] and the word ʿApiru occur together 125

1. See the works of Moshe Greenberg, *The Ḫab/piru*. and J. Bottéro, *Le Problème des Ḫabiru, Cahiers de la Société Asiatique*, no. 12 (Paris, 1954).

2. For an extensive discussion with bibliography, cf. H. H. Rowley, *From Joseph to Joshua* (London, 1950), pp. 37–56. See also *LIS*.

3. Greenberg, *The Ḫab/piru*, pp. 87, 92.

4. The term "ideogram" is deliberately used here on the ground that we cannot prove that the writer intended any specific word (thus "logogram"). To use the term is to beg the question, particularly in the case of borrowed writing systems. I maintain that the written sign frequently refers directly to the historic reality involved—that it does not act as an exact substitute for a spoken word. One wonders how the Egyptian court scribes read it! Cf. Greenberg, *The Ḫab/piru*, p. 86, n. 1.

times in the Amarna letters. The distribution of occurrences would seem to the present writer to preclude the possibility of locating the *SA. GAZ* in any particular region of Palestine or Syria; much less is it possible to show an invasion route followed by a migrating horde. The ᶜ*Apiru* are mentioned as attacking Byblos and controlling Ṣumur on the coastal plain,[5] present at or near Qedeš and Damascus,[6] raiding villages in the Biqᶜa,[7] and in the south they are mentioned in nearly every corpus of correspondence except that of Widia of Ascalon. There are only six corpora of letters which lack the term (excluding the international diplomatic correspondence from Hatti, Mitanni, Assur, and Babylon). It is highly significant that both Abdu-Ashirta and Aziru never use the term. Furthermore, it does not have the same connotations in the various collections, and if it does not refer to an ethnic stock, the sociopolitical meaning of the term would vary somewhat since the letters come from a great variety of cultural and linguistic backgrounds.

Table 1   Distribution of Occurrences of *SA. GAZ*/ᶜ*Apiru*

Occurrences			Lacking		
Ruler	*City-State*	*No. of Occurrences*	*Ruler*	*City-State*	*No. of Letters*
Rib-Addi	Byblos	45	Ammunira	Beirut	3
Zimreda	Sidon	2	Widia	Ascalon	10
Abimilki	Tyre	2	Abdu-Ashirta	Tunip?	6
Aitugama	Qedeš	3	Aziru	Tunip	11
Mayarzanu	Hazi	35	Akizzi	Qaṭna	4
Biryawaza	Ube	4	Addu-*UR. SAG*	unknown	2
Biridiya	Megiddo	2			
Labayu	Shechem	1			
Milkilu and Iapaḫi	Gezer	6			
Abdu-Heba	Jerusalem	10			
	unknown	9			
Shuwardata	Keilah	1			

The chart shows at least that there is no necessary correlation between geographical location and use of the term. Ammunira of Beirut complains of the same hostility of which Rib-Addi became a victim, but does not mention the *SA. GAZ*. The same contrast is to be found in the cases of Qaṭna and Qedeš, Ascalon and Gezer. Many details are of course lacking, but the argument from silence which this presents at least fits in with the

5. *EA* 68:12–14; 74:14–22; 67:16–18; 68:14–18.
6. *EA* 197:26–34; 189:9–12; 195:27.
7. *EA* 185, 186.

prevailing opinion that there was no ethnic stock involved in the *SA.GAZ* movement, but that rather the term was an appellative which might or might not be used depending upon the subjective opinion of the writer. It was not a designation by which the hostile groups involved identified themselves, nor were the groups so designated set apart clearly from other elements in the population of Palestine and Syria in every case.

In only two passages in the whole of the Amarna letters can a *SA.GAZ* be identified by name. There seems to be a curious reluctance to apply the term specifically to an individual, and it is usually treated as a collective with the determinative *LÚ.MEŠ*, even when, as in these two passages, it does designate an individual. The two passages in question are *EA* 91:3ff and 67:17ff, both unfortunately in broken condition. In the first, following Knudtzon, we read:

[*a–n*]*a m*[*i–ni*] *aš-ba-ta* [*ù*] *qa*[*-l*]*a*[*-t*]*a ù yi-il-qú*[*alāni-k*]*a* ᴸᵘ*GAZ*ᵐᵉˢ *kalbu* [*i-nu-ma*ᵃ]ᴵ*Su-mu-ra yi-il-qa* [*ù aš-t*]*a-par a-n*[*a k*]*a-tú a-na mi-ni* [*qa-la-t*]*a ù t*[*u*]*-ul-qí* [ᵃ]ᴵ*B*[*ít-*]*ar-q*[*a i–n*]*u-ma yi-d*[*a–*]*gal* [*ù*] *ia-nu* [*š*]*a yi-iq-bi m*[*i-*]*a*[*m*]*-ma* [*a-na ša-š*]*u muḫḫi* ᵃᴵ*Su-mu-ra*
"Why have you sat [back] and done nothing so that the *GAZ* [pl!], the dog, takes your cities? When he took Ṣumur I wrote to you, 'Why have you done nothing?' and Bit-Arqa was taken when he saw that there was no one who said anything to him because of Ṣumur."[8]

There can be no reasonable doubt that it is Abdu-Ashirta who is here called the *GAZ*; it was at his command that Bit-Arḫa was taken. He took Ṣumur. *Kalbu* 'dog' is Rib-Addi's favorite epithet for Abdu-Ashirta (and for his son).[9]

The second passage refers most likely to Aziru:

i-na-an-na šu-ú-tú k[*i-ma* ᴸᵘ]*SA.GAZ.ZA*ᵐᵉˢ*kalbu ḫal-qú ù iṣ-ba-at*[ᵃᴵ*Su-*]*m*[*u-r*]*i*ᵏⁱ
"Now he is like a *SA.GAZ*, a fugitive dog, and has seized Ṣumur."[10]

These are the only two passages in which an identification of a *SA.GAZ* is certain, or reasonably so. There are many parallels between Abdu-Ashirta and Aziru on the one hand and the *SA.GAZ* on the other, which effectively equate the two in Rib-Addi's subjective opinions. In his general descriptions of what is happening to the empire, Rib-Addi uses four generalizations repeatedly:

8. This vitiates part of William Moran's treatment of *EA* 74 ("Amarna *šumma* in Main Clauses," *JCS* 7 [1953]: 78–80). To "play by ear," so to speak, is the whole point of the clever political policy of Abdu-Ashirta. The failure of the Egyptian administration to call him to account for one venture constitutes encouragement to take the next step.
9. *EA* 79:21–22; 84:11–16; 84:17, 34; 85:64; 125:40–41.
10. *EA* 67:17ff. The writer of the letter has not been identified with certainty.

1. "The *GAZ* have taken all the lands" (83: 16ff.; cf. 90: 25; 117: 58).
2. "Abdu-Ashirta (or Aziru) has taken the land(s) of the king" (71: 18–19, 28-32; 74: 23-24; and 24 other passages).
3. "The cities/lands will join/have joined the *GAZ*" (73: 26ff., 32f; 74: 19ff, 27, and 13 other passages).
4. "The cities/lands will join/have joined Abdu-Ashirta/Aziru" (88: 31–32; 79: 42ff.; 84: 9–10, 12–13; 138: 93; 74: 27–29).

The frequency of these statements and their apparent interchangeability even within the same sentence[11] make it clear that the hostility of Abdu-Ashirta and that of the *GAZ* are not two different things for Rib-Addi, but merely different ways of saying the same thing. The relation between the two cannot, of course, be one of complete equation, because the *GAZ* are involved in areas outside the realm of activities of Abdu-Ashirta and his sons, at least so far as we know. The explanation suggested in the past[12]— actually based on an ethnic interpretation of the term *SA.GAZ* which saw them as invaders with whom Abdu-Ashirta made common cause in order to use them for his own personal ends—founders on the following facts: (1) There is no hint that the movements of Abdu-Ashirta and the *SA.GAZ* are two different things, so that there can be any activity of the *GAZ* in the area in question which does not directly involve Abdu-Ashirta. (2) In those descriptions of hostility sufficiently detailed for us to draw conclusions, the attackers are the populations and heads of city-states, the latter—even Abdu-Ashirta and Azira—being men who were placed in their positions by the Egyptians themselves.[13] Rib-Addi describes very clearly what was happening: "And the people defect in order to take the land for themselves, and there is no one to guard Byblos, the city of the king, my lord."[14] He says of the peril of Byblos,

Hostility is powerful against me, and there are no provisions for the *ḫupšu*, and [therefore] they defect to the sons of Abdu-Ashirta, and to Sidon and to Beirut. The sons of Abdu-Ashirta are hostile to the king, and the city of Sidon and [!] the city

11. *EA* 88:29–34.
12. Cf. Rowley, *From Joseph to Joshua*, and literature there cited, for the old discussion of the view that the Amarna Letters are in some way connected with the biblical account of the Israelite conquest of Palestine. Also *LIS*, pp. 66–102.
13. *EA* 101:29–31; 161:51–53.
14. *TET*, 129a:35–39. "Desertion" usually means the rejection of Egyptian authority, not necessarily the movement from one place to another. The Amarna word *paṭāru* corresponds semantically to the Hebrew word *sûr*, and can be used for abstractions; *derek* and *miṣwâ* are used after *min* in Hebrew, exactly as *awātu* is used after *ištu* in Amarna with the verb *paṭāru*. Cf. *EA* 166:18. Further parallels need not be cited here, but include also prepositional phrases indicating the person who is forsaken (Yahweh in Hebrew; the king in Amarna). I will not venture a guess as to the Canaanite word that dictates the semantic range of the Amarna word *paṭāru*.

of Beirut do not belong to the king. Send a commissioner, let him take them! Let not one city be left [i.e., outside Egyptian control] lest it desert you. Behold, when the *ḫupšu* defect then the *GAZ* seize the city.[15]

Internal power struggles in the city-states, not an invading horde of foreign barbarians, results in *GAZ* control. It is the *defection* of Ṣumur and Bit-Arḫa which resulted in the control of Abdu-Ashirta and the *GAZ*.[16] In the same letter Rib-Addi threatens to make a covenant with Abdu-Ashirta like that of Tyre and Sidon if the king does not respond to his appeals for help. (3) Several letters actually refer to cities *becoming SA.GAZ*. That is, in the opinion of the letter writer, they have defected from their political allegiance by joining a conspiracy-revolt movement.

It is particularly striking that although Rib-Addi complains ad nauseam of the hostility against him he is able to actually specify almost no clear case of overt violence performed by the "*GAZ*" or by Adbu-Ashirta and his sons. He uses the words *laqû*, *ṣabātu*, *nadān pāni* of the *GAZ* perhaps deliberately in the hope that the pharaoh will be misled, as modern scholars have been, into thinking that there was repeated storming and burning of cities. He mentions the burning of cities but twice—once in reference to the Hittites, of whom he "had heard" that this was their practice,[17] the other time simply as a general protest against the high-handed actions of the sons of Abdu-Ashirta.[18] It is no wonder that in reply Abdu-Ashirta and his son both complain that they are being slandered.[19]

If we disregard these generalized accusations of Rib-Addi, it becomes clear from his inability to cite specific actions of violence that the situation was not one in which invading hordes were storming and seizing cities by force, but simply one of opposing political groups and possibly programs.[20] There can be no doubt that the political orientation and control of such cities as Ṣumur, Bit-Arḫa, Ammiya, and others (including finally Byblos itself) did change during this period, but the change was not of a sort that convinced the pharoah that they were lost to the empire. It is this which Rib-Addi presents as his major thesis—and this is the meaning of *SA.GAZ* as Rib-Addi uses the term.[21] It is certain that the Egyptian commissioners were unable

15. *EA* 118:21–39.
16. *EA* 83:28–29.
17. *EA* 126:51–52 (cf. 55:41).
18. *EA* 125:40–45.
19. *EA* 62:39–45; 161:7–9.
20. The obvious political "program" is that described in *EA* 74. Cf. Mendenhall, "The Message of Abdu-Ashirta to the Warriors, *EA* 74," *JNES* 6 (1947): 123–24, and Moran's corrections (see n. 8, above).
21. It is necessary to reemphasize that there is no reason to assume that the term has the same meaning in every corpus of correspondence in the Amarna Letters, and there is definite evidence to the contrary (see below).

to preserve the peace between the various regents. The Egyptian king evidently cared little about which city was in whose hands so long as the kinglet in question continued to pay his tribute and give assurance of his undying loyalty. Therefore, Rib-Addi had to assume the burden of proof in order to convince the pharoah that the interests of the Egyptian Empire demanded that Rib-Addi be supported against those who were wresting his cities from his control. Rib-Addi took the stand that a city lost to him, since it had presumably been put in his charge by Egypt, was actually lost to Egypt. The king, on the other hand, who received both sides of the argument, was unconvinced (together with his officials). Indeed the pharaoh saw little point in undertaking expensive campaigns in Palestine and Syria to support a weakling who could not control his own territory, when a strong and dynamic leader such as Abdu-Ashirta or his sons could convince the Egyptians that he could maintain strong control of the lands.

The unfortunate Rib-Addi, deprived of much of his territory, could not even support his own *ḫupšu*[22] 'peasants,' much less send tribute enough to impress the increasingly parsimonious Egyptian court.[23] In such a situation, his Cassandra-like portrayals of calamity took the form of accusations against seemingly faithful vassals of Egypt, which must have appeared completely irresponsible to the Egyptian officials.

THE ROLE OF THE *SA.GAZ* IN THE COLLAPSE OF THE EMPIRE

The thesis to be defended here is that the *SA.GAZ* were not the cause of the collapse of the empire in the sense that an invading group captured the various cities under Egyptian rule by military action. Rather, they represent a movement of political nature which arose because of the inability of the Egyptian officials to maintain order among the various city-states in the empire. In order to discuss the meaning of the term as it is used by Rib-Addi, we must first reconstruct the process by which Abdu-Ashirta and Aziru gained control. Rib-Addi's greatest justification for his use of the term consisted in the fact that seizure of control of territory and cities did not take place by normal legal and administrative processes. There can be no reasonable doubt that his enemies did succeed in gaining control of cities in violation of the previous commitments of the empire,[24] but they did

---

22. *EA* 85:8–11.

23. Cf. *FSAC*, pp. 206–9 for a general picture of the empire. Note also the indignant reproach of Kadašman-Ḫarbe for the debasement of the gold standard. *EA* 3:13–22; 7:66–72. Cf. Tušratta's complaint (29:69–72) about gold-plated wood!

24. *EA* 62, where Abdu-Ashirta offers an elaborate explanation of his seizure of Ṣumur.

so by political intrigue of such a nature as to deprive Rib-Addi of any really convincing grounds for his complaints rather than primarily by military action. Consequently, he was unable to gain redress for his grievances through the normal administrative processes of the empire. Certainly he was not the only regent to suffer thus from the illegal acts of ambitious politicians, and his greatest handicap was evidently his unwillingness or inability to engage in the rough and tumble game of political self-help counter to the legal obligations imposed by the empire on its vassals, probably by covenant.

When law and order broke down, the strongest and cleverest could expand their control, maintaining just enough outward show of loyalty to Egypt to stave off the danger of an Egyptian military expedition. There can hardly be any doubt that northern Syria in particular accepted help from Mitanni or Hatti,[25] not for the purpose of exchanging loyalties, but in the vain hope of maintaining independence by playing one power against another. This seems to have been an increasingly important factor at the time of the Amarna letters themselves. Rib-Addi describes the strength of Abdu-Ashirta as deriving from the *GAZ*; instead, it is clear that Aziru, son of the latter, was much involved with the Hittites.[26] The process of wresting control from Rib-Addi can be reconstructed from the data he gives as follows:

1. Elements in the population of a given city are won over, becoming adherents of Abdu-Ashirta. The best example is Byblos itself.[27]
2. Those sympathizers in the city are encouraged to assassinate the regent loyal to Egypt as a means of freeing themselves to "join" Abdu-Ashirta or his "sons." This happened in Byblos itself,[28] and in Ambi[29] and Šigata. Similar procedures are attested at various cities in the south of Palestine, probably at a later time.[30]
3. A regime faithful to Abdu-Ashirta is placed over the city (in Byblos, Ambi, and Šigata).[31]

25. *EA* 126:59; 53:8–10. Cf. also the broken passage (75:41–42).

26. *EA* 71:20–22; 126:58ff.

27. *EA* 138:71–73 (Aziru).

28. *EA* 139:38. An attempt on the life of Rib-Addi at the command of Abdu-Ashirta is described in 81:14–19. Cf. 82:37–41.

29. *EA* 74:25–29. Cf. 140:11, where Aziru is accused of the same crime.

30. *EA* 287:71–75; 250:16–18. It goes without saying that any political assassination in this correspondence will be construed as a hostile act against the king of Egypt in order to gain the support of the imperial power. Cf. also the assassination in Tyre (89:20–21).

31. *EA* 104:51–52; 138:59–63; Ambi: 76:18–20; Šigata: 74:24; 76:18–20. Note the title *bêl ālim* 90:27–28; 102:22–23; and cf. *baᶜalê* followed by a place name in Hebrew (Josh. 24:11; Judg. 9:2; 1 Sam. 23:11–12).

4. An alternative to steps 2 and 3 is the winning over of the regent himself, who would understandably be willing to save his own skin. Rib-Addi was sorely tempted, and probably did capitulate after his exclusion from Byblos.[32] This usually took place by covenant. Aziru had a covenant with Sidon and Arwad which gained for him control of all the approaches to Ṣumur.[33]

5. Economic and military pressures or threats are used to induce other cities to join the movement. Sidon and Tyre held out for a time as well as Byblos. This may very well have been a most useful device to insure the disaffection of the masses within a city. The population of Byblos was disaffected by the fact that its food supply was cut off, and Tyre is described as suffering from a similar situation in the letters of Abimilki.[34] The process involved harrassment of the peasants who therefore could not till their fields or harvest their crops.[35]

6. The final goal was the unification of the land by covenant, if Rib-Addi's source of information is to be trusted.[36] This was never achieved so far as we know, though Aziru must have made considerable progress toward the goal. The intervention of Hittite power plucked with little difficulty the fruits of the *SA.GAZ* movement.[37]

The invasion of nomadic bands is completely irrelevant to the whole process, even pastoralism, and is never clearly referred to in the whole corpus of Amarna letters, in striking contrast to the Mari letters, where the tribal pastoralists are frequently referred to and are the objects of a very close surveillance.[38] Only once is the term *SA.GAZ* the subject of a verb indicating movement, and in that one case it is a movement of troops from Ṣumur to the vicinity of Byblos.[39] Not once does Rib-Addi mention the capture of a city by battle—it is always implied as an immediate danger.[40]

32. *EA* 67:13–14? The letter is too badly broken to give us the precise context (cf. 83:23–25). In the days of Aziru: 125:39–40; 138:44–50; 162:2–5, 18.

33. *EA* 149:57–60. Cf. also Abdu-Ashirta's treaty with Iapa-Addi of unknown location and Zimreda of Sidon (83:24–26).

34. *EA* 85:8–15 *et passim*, and 148:9–13, 30–34; 149:51–53; 154:11–18. The issues are water supply, wood, and burial grounds.

35. Cf. the same policy in the south (*EA* 244:11–16).

36. *EA* 74. Cf. n. 20.

37. For the later history of the area controlled by Abdu-Ashirta and Aziru, see O. R. Gurney, *The Hittites* (London, 1952), pp. 74–75.

38. Luke, "Pastoralism and Politics." Cf. *PSO*, p. 146.

39. *EA* 108:62–65. Context broken and obscure.

40. *EA* 75. Cf. *EA* 244 from Megiddo in the south. The process is not one of pitched battles, but rather of harrassment, psychological warfare, assassination, and political intrigue. Cf. *PSO*, p. 115.

Abdu-Ashirta and Aziru were much too astute as politicians to commit an overt act that could serve as incontrovertible evidence against them. The raids that seem to have taken place against open country were evidently a more or less normal state of affairs at this time, which in the mind of Rib-Addi served to heighten the similarity between Aziru/Abdu-Ashirta and the typical sort of *SA.GAZ* band.[41]

ABDU-ASHIRTA/AZIRU AND THE *SA.GAZ*

The main argument in the letters of Rib-Addi is that territory lost to Abdu-Ashirta and Aziru is territory lost to the empire. This is also his main intention in his use of the term *SA.GAZ*. In *EA* 67:17 (not from Rib-Addi) we have an actual definition of the term. *Kalbu ḫalqu* is in apposition to *SA.GAZ*, defining the latter by analogy. A stray dog is one which has become alienated and is beyond the control of his master. In other words, all that the word demands, as Rib-Addi uses it, is that a relationship which had existed is now broken when a person or group becomes *SA.GAZ*.[42] When a regent is a faithful vassal, he is an obedient dog, "hearing his master's voice."[43] Having become a *SA.GAZ*, he is a stray dog who pays no attention to the representatives of the king nor to his normal obligations as a vassal. The term is thus, for Rib-Addi, an abstraction, full of emotional and political bias like the term "communist" in modern American politics, by the use of which Rib-Addi hoped to gain sympathy for his cause. Even Rib-Addi had to admit tacitly that his opponents were not actually a group to which this designation was ordinarily attached, but rather that their actions were *similar* to those of *ᶜApiru* bands.

41. Cf. the complaint of the Cassite king (*EA* 8:13–34) against acts of Egyptian vassals.

42. *EA* 74:19–21, 27–29. If the position taken here is correct, then all the speculation as to the etymology and original or basic meaning of the term misses the point. That is to say, if the term applied only to the *fact* of a broken relationship, the manifold variants of the social conditions attested for the *ᶜApiru* by our sources are completely irrelevant, and no conclusions can be drawn from the fact that these social conditions may have something in common. It must be conceded, however, that the important concepts emphasized in Rib-Addi's letters cannot be proven to be constants in all occurrences. What is striking, however, is the fact that much evidence shows that *ᶜApiru* groups were frequently displaced persons, persons not in their original social context (cf. Greenberg, *The Ḫab/piru*, *passim*, especially pp. 61–96). What happened after the initial breach is irrelevant to the fact that the breach had taken place. The most obvious procedure would be, first, the adherence to others in a similar situation, and, second, the group's adherence as a whole to some other political authority—without, however, losing their identity as an *ᶜApiru* group, or becoming "citizens" in their new political context.

43. *EA* 328:21–26 *et passim*; cf. Exod. 24:3.

Though admittedly the procedure is somewhat precarious, it should be possible to reconstruct the normal usage of the term from the extended or figurative use in the Rib-Addi correspondence. The hypothesis here presented would suggest that the meaning is derived from a distinction between an in-group and out-group, based not on ethnic, folk, or kinship ties (in Palestine and Syria such ties must have long since ceased to be the effective basis of *political* organization by the time of the Amarna corpus) but rather on political and legal basis. If a *SA.GAZ* were a stray dog, he would be one from the point of reference of a political or legal organization to which he had formerly belonged. In Rib-Addi's letters that point of reference is of course the Egyptian Empire, so that the *SA.GAZ* of whom he speaks are such simply because they have, in his opinion, illegally broken their relationship to the king. To summarize, it is here suggested that the meaning of *SA.GAZ* would include three basic and constitutive factors, all of which have been transferred from the normal social meaning of the term to serve Rib-Addi in his political polemic against his adversaries.

1. *The loss of status.* Tribal-pastoral groups are not demonstrably *SA.GAZ* as such in any of ancient oriental literature.[44] They have their own recognized type of social organization and confer status upon their own members. Only if a member were ejected or fled from the group might he be termed *SA.GAZ*—he is like Cain. So also, Rib-Addi repeatedly states that cities and territories which are under the control of the *SA.GAZ* are not actually a part of the king's possessions any longer. When they join Abdu-Ashirta, similarly, they are no longer under the king. This is Rib-Addi's "subjective opinion," however, which is countered by the affirmations of undying loyalty to the king of Abdu-Ashirta and his successors. At any rate, it is clear that to Rib-Addi *SA.GAZ* groups are no longer under the control of the king.[45]

2. *Twofold nature of loss of status in a legal/political community.*

a. The group so designated does not feel bound by the legal obligations of the community, and is not controlled by that community. This again is emphasized by Rib-Addi. Abdu-Ashirta and Aziru "do what they please" in the lands of the king, as though they were independent "great kings" like those of Mitanni and Hatti.[46] They are taking the lands of the king for

44. Cf. J. Wilson, in *ANET*, p. 247, n. 47, where they are clearly distinguished from normal geographical designations.

45. *EA* 88:29–34 *et passim*.

46. *EA* 76:14–16; 104:17–24; cf. 109:9–11. The parallel to Judg. 21:25 in meaning though not in language is striking, and perhaps not coincidental. After the establishment of the Monarchy there could hardly have been any similarity felt between the Israelites and the old ᶜApiru groups. During the time of the Judges, on the other hand, the similarity must have been striking from the point of view of their neighbors. Cf. Num. 23:9.

themselves. The letter of Amenophis III (?) to Aziru indicates that there is much truth in Rib-Addi's contention, though Aziru is not reprimanded for the actions of which Rib-Addi complains. Most serious is Aziru's establishment of covenant with an enemy of the king, in flagrant violation of the customary prohibition of a vassal's making covenants with another foreign group.[47] At any rate, it is certain that Aziru did not feel himself bound by the customary obligations of a vassal.

b. If status in a legal community imposes obligations, it also confers protection and privileges. The individual is exempted from the danger of attack by superior force, both within and without. The authority of the community over him is exercised by legal, orderly procedures. Here again, the SA.GAZ are a group apart, since they cannot be dealt with by legal procedure, but only by military force.[48] It is within this framework that Rib-Addi's letters gain added meaning—nearly every letter concludes with an appeal to the king to send his *armies* to reestablish control. Since the Egyptian court did not agree with Rib-Addi's analysis of the situation, it is not difficult to understand why, after the single expedition associated with the demise of Abdu-Ashirta,[49] they sent no armies to the field. This appeal for force would be unthinkable, as Rib-Addi himself points out, had the maintenance of law and order not, in his opinion, completely broken down. Rib-Addi's appeal is to the executive aspect of government since the judicial aspect could not cope with a situation where it was not recognized.[50]

3. *The illegitimate use of force.* Exercise of force, power, is a function of legitimate government. Legitimacy is a constant concern of governments and kings, founded ultimately upon a concept of delegation of authority

---

The historic implications of this may be far-reaching. The majority of the Israelite tribes must then be regarded as having had a political obligation to other political structures prior to their adherence to the Israelite federation. This is a priori entirely likely, in which case the Amarna Letters are not entirely irrelevant to the history of the conquest. It is not the same movement, historically; it is rather a very similar one that may have proceeded in similar fashion. The difference would lie in the fact that the Canaanite city-states saw important differences that actually led to war—for which we have practically no evidence in the Amarna Letters.

47. *EA* 162:22–27.

48. *EA* 144:24–32.

49. *EA* 138:28–34. Other references to Egyptian military expeditions are numerous —e.g., 337:13–18, but we have no historical context. It could have been the same occasion as that referred to in *EA* 138—or any later foray.

50. The Amarna corpus makes it difficult to prove that the Egyptian suzerainty was sealed by a formal covenant of any sort; i.e., no sworn oath is mentioned. Breach of covenant is not mentioned in the Rib-Addi letters (or, for that matter, elsewhere). Contrast Mari (*ARM*(T), I, 3:5–9). Here again, the biblical tradition contrasts with that of the Canaanites as we know it from the Amarna Letters, and shows its intimate connections with North Syrian traditions.

from the divine world.[51] This religious aspect of illegitimacy of the *SA.GAZ* movement is not dealt with directly by Rib-Addi, but *EA* letter 74 gives us a hint of Abdu-Ashirta's religious ideology. The fact that the coalition was to be formed by an assembly at *Bît-NIN.IB*, the temple of some god, rather than at one of the more frequently mentioned cities controlled by Abdu-Ashirta indicates some sort of religious sanction underlying the proposed coalition.[52] Again, characteristic of several ʿApiru groups, as attested in western Asiatic sources, is an established relationship between the group and a deity, usually through the mediation of a priest. The relationship may exist for the purpose of gaining the aid of the divine world in maintaining the precarious existence of an outlaw group rather than for establishing legitimacy of a government. So Idri-MI acted as a consultor of omens—in a priestly capacity—for the ʿApiru of Ammia;[53] an Alalakh ʿApiru group included a priest of Išḫara;[54] and David used the services of Abiathar to consult the will of the Yahweh in his attempts to escape from Saul.[55] The gods of the ʿApiru mentioned in the Hittite treaties may refer to this characteristic of ʿApiru groups, these gods themselves being bound to refrain from aiding renegades who had fallen out with the Hittite Empire, whether or not they were actually regarded as "patron deities" of the group.[56]

Rib-Addi repeatedly questions the legitimacy of Abdu-Ashirta's actions.

Who is Abdu-Ashirta, the slave, the cur, that he takes the land of the king for himself?[57]

Who is he, Abdu-Ashirta, the cur, that he seeks to take all the cities of the king, the sun-god, for himself? Is he the king of Mitanni or the king of Kaššu that

51. Cf. E. A. Speiser, "Authority and Law in Mesopotamia," *JAOS* suppl. no. 17 (1954): 8–11.

52. Cf. V. Korošec, *Hethitische Staatsverträge* (Leipzig, 1931), pp. 92ff., on religious sanctions in covenant relationships.

53. S. Smith, *The Statue of Idrimi* (London, 1949); W. F. Albright, "Some Important Recent Discoveries: Alphabetic origins and the Idrimi Statue," *BASOR* 118 (1950): 14–20.

54. *AT*, p. 71, no. 180; Bottéro, *Le Problème des Habiru*, p. 183.

55. 1 Sam. 23:6–14.

56. Cf. Greenberg, *The Ḫab/piru*, pp. 76–78. Since the purpose of the list of witnesses to the covenant was to obtain the sanctions of the divine world, the inclusion of a divine name in the list does not necessarily imply anything with regard to the status of the deity in Hittite society. Cf. Hillers on the "former gods" in a forthcoming article; H. G. Güterbock, "The Composition of Hittite Prayers to the Sun," *JAOS* 78 (1958): 239–40. (This may well give the context or contrast for the "new gods" of Judg. 5:8—which led to war.) That any supernatural aid can be forthcoming to anyone who becomes guilty of breach of covenant is thus denied.

57. *EA* 71:16–19.

he seeks to take the land of the king for himself? Behold now he has mobilized all the *GAZ* against Šigata and Ambi, and has taken these two cities.[58]

Who are the curs, the sons of Abdu-Ashirta that they do what they please, and burn the cities of the king with fire?[59]

It is difficult to illustrate this aspect of the argument in more detail, since *SA.GAZ* occurs as the subject of a verb only thirteen times in the Amarna letters. In every case, however, the verb used implies illegitimate seizure or violent action against the property of the king—usually, moreover, as a vague generalization. It seems likely that the *GAZ* acted violently—that is, by the exercise of private force in violation of the obligations imposed by the legal community from which they originated.[60] The question is whether or not anyone who indulged in illegal self-help might therefore have been called a *SA.GAZ*. It would be impossible and a priori unlikely to prove this, for every legal community tends to recognize at least certain situations in which the exercise of extralegal self-help would not necessarily bring about those sanctions which would exclude the actor from status in the group—whether by voluntary exile, banishment, imprisonment, or death.[61] It is being argued here merely that the illegitimate use of private force against the demands of the existing political structure is one of several characteristics of a *SA.GAZ* group as Rib-Addi uses the term. He emphasizes this characteristic of Abdu-Ashirta in order to convince the Egyptian court that Abdu-Ashirta is not to be regarded as a person with legal status who can be adequately dealt with by normal political channels.

Thus far our main contention has been that Rib-Addi applied a term to his enemies that did not really fit—that he really used an abstraction based on his subjective view that the *SA.GAZ* and his enemies had much in common. The term was doubtless applied to specific, historical groups as well as to individuals, but opinions as to what these common factors might be will differ. Those factors can hardly have been much different, however, for Rib-Addi was concerned with making as strong a case as possible. It

58. *EA* 76:11–20.

59. *EA* 125:40–45; cf. n. 46.

60. Greenberg, *The Hab/piru*, p. 95, but cf. p. 76. The difference between the violent and nonviolent *SA.GAZ* is a function of their social relationships (or lack of them) after the initial breach. See n. 42, above. After entering the service of Achish, king of Gath, was David a violent or a nonviolent *ʿApiru*? Violence can be accepted in the ancient world as a function of all social groups. Once an *ʿApiru* group entered the service of a recognized political entity, its activities were presumably controlled—Achish thought so too.

61. Compare the frequent refusal of grand juries in modern society to indict in cases of self-help exercised by outraged husbands. Cf. Prov. 6:34—the only example other than Gen. 4:24 where the root *nāqam* is used to refer to human action without pejorative connotation. See chap. III.

would hardly have seemed adequate to him to say "any enemy of mine is a *SA.GAZ*"—this is somewhat too modern an argument. I suggest that he uses the term of his enemies because of a direct, intuitive, and emotional perception of a similarity between Abdu-Ashirta/Aziru on the one hand and *SA.GAZ* groups on the other; what the similarity consisted of we have attempted to describe in the foregoing pages.

Real *SA.GAZ* groups also appear in these letters, however. In *EA* letters 185 and 186 we have descriptions of the actions of such a group. Here there are complaints of the illegitimate use of force. The *SA.GAZ* are engaging in repeated raids against neighboring villages, attacking, looting, burning, and then returning with the booty to their base of operations. This action brings about a legal claim against Amanḥatbi, their *patron*. Though the details are not clear, the plaintiffs could not obtain redress from Amanḥatbi (obtaining redress from the *SA.GAZ* directly by legal means is not even mentioned), and so, appeal directly to the king. However, the plaintiffs do not declare war on Amanḥatbi, evidently because they have confidence in the regular procedures of the empire, and they also emphasize the fact that they are faithful vassals of the king.[62]

In our opinion, the only nontypical and therefore complicating factor in this incident is the fact that the *SA.GAZ* group in question has some sort of established relationship to Amanḥatbi, who in turn is evidently a regent of the empire. Biryawaza of Damascus also casually mentions "my *SA.GAZ*" without any indication that there is anything abnormal or illegitimate about the relationship. Evidence elsewhere also indicates frequent use of *SA.GAZ* groups as mercenaries in the service of a king. Use of a group of mercenaries does not however mean that they are "citizens" of the state which they serve. Consequently, any actions against them must take the form of action against the person whom they serve, and who, therefore, has legal responsibility for them.[63]

ᶜAPIRU AND HEBREWS

The final point to be raised in this chapter is a defense of the equation of ᶜApiru and Hebrew on this nonethnic but legal and political ground. In biblical traditions there are repeated examples of the sort of phenomena associated with the Amarna *GAZ*. The clearest example is that of David. He lost status in the Israelite community by flight caused by the enmity of the king. There gathered about him other refugees motivated by economic

62. *EA* 185, 186.
63. Cf. n. 42, above.

as well as other concerns.[64] All were similarly without legal protection, and had to maintain themselves by forming a band under the leadership of David, which was then able to survive by cleverness combined with a considerable degree of mobility. The temptations of exercising self-help in violation of the moral obligations of the Israelite religious federation are vividly (and deliberately?) described in chapters 24 (= 26) and 25 of 2 Samuel.[65] As any such band must have done when governmental control was fairly well established, they had to establish a relationship with a foreign power, Achish of Gath, for survival. They continued to live by raids directed against the enemies of Israel (double-crossing Achish, just as the Amarna ʿApiru may have double-crossed Amanḫatbi by attacking those with whom he had legal and political relationships). Finally, the Philistines themselves label David and his gang ʿApiru—"Hebrew," and relieve David of the embarrassing moral choice of open breach of either his political obligations to Achish or his religious obligations to Israel, connection with which he never renounced[66] (in spite of his reproaching Saul with having "driven him out to serve other gods").

The story of Jephthah is very similar—deprivation of legal status because of illegitimate birth (and no doubt, unwillingness of his half-brothers to share the paternal inheritance—in violation of Israelite customary law?[67]) and reinstatement as chief because of his military power. The fact that the term ʿIvrī is not used does not obscure the fact that the pattern is the same.[68]

It has already been pointed out,[69] but insufficiently taken into account,

64. The similarity has been noted elsewhere. See Albrecht Alt, "Erwägungen über die Landnahme der Israeliten," in *Kleine Schriften*, 1 (Leipzig, 1953): 170, and especially n. 3: "Die dort folgenden Erzählungen geben ein gutes Bild von der wechselnden Verwendung einer solchen Freischar bis zu ihrer Überführung in den Dienst eines fremden Herrschers." The subsequent history or occupation of an ʿApiru group has nothing to do with the factors that constitute it.

65. These two illustrations shed vivid light upon the moral and ethical nature of early Israelite religion. Further implications of these episodes are discussed in chap. III.

66. This is illustrated by his refusal to reject the religious obligations noted above. His reproach of Saul is based upon what is *normal* in ancient times; we have every indication in the narrative that he consistently considered himself a worshipper of Yahweh. His adherence to the moral norms of Yahwism are in keeping with this, and also illustrate the fact that obedience to ethical norms was at least as constitutive of Yahwism at this time as was any official cultus from which he was naturally excluded by virtue of his status as ʿApiru. With this, then, compare the reported words of Samuel (1 Sam. 15:22–23).

67. Can we assume that Gen. 21:10–11 and the legal custom presupposed there was still normative in the time of the Judges?

68. Greenberg, *The Ḫab/piru*, p. 76, n. 73.

69. E. A. Speiser, "Ethnic Movements in the Near East in the Second Millennium B.C.," *AASOR* 13 (1933): 43; Greenberg, *The Ḫab/piru*, pp. 92–93.

that the patriarchal narratives indicate a similar pattern of social behavior. Abraham and Jacob, as well as others in the patriarchal period, are chieftains not of nomadic tribes but of ʿApiru groups which have no legal status and have indeed severed themselves from an earlier political community. Finally, the late confession in Deuteronomy 26:5 presents the same picture. ʾArammī ʾōvēd has the same meaning as kalbu ḫalqu in Amarna. Jacob was neither a "wandering Aramean" nor a "Syrian ready to perish." He was in fact a "fugitive Aramean" who by flight and stealth cut himself off from the community of which he had been a member, and, according to biblical tradition, was rescued from extermination only by the direct intervention of Yahweh.[70]

The designation ʿIvrī for the Israelites is then the last preserved usage of a term which had applied to any number of stateless persons and groups in the second millennium B.C.[71] It came to be applied to Israel because there was a continuity in pre-Israelite tradition and history of refusal by villagers and shepherds to become assimilated to the existing political organizations in whose environs they lived.[72] When the political empire became intolerable and unable to preserve order, they withdrew from *all* obligation and relationship to it[73] in favor of another nonpolitical overlord (whose obligations were of an entirely different and functional order). This was what being an ʿApiru meant in early Israelite times. It was only under the Monarchy that they ceased to be ʿApiru and became a nation. Consequently, the term ʿApiru ceased to be a politicolegal term and became an "ethnic" designation in addition to the proper name for the religious federation, Bᵉnē Yiśrāʾēl.[74] The ʿApiru whom the new state created by its intolerability had to be designated by some other term, for none existed in the time of David's

---

70. Greenberg, *The Ḥab/piru*, p. 96. The formula can hardly be earlier than the tenth century.

71. The impressive fact that the sources (except for the Bible) cease to use the term after ca. 1200 B.C. is best explained by the fact that the great empires of the Late Bronze Age, and the well-defined political and legal relationships that accompanied them, disappeared in the general chaos that followed. Consequently, such a term would cease to have meaning.

72. E.g., Jacob, whose wily outwitting of Laban is then paralleled by the wily removal of the Moabite king Ehud, the strategems of Gideon, of David, and of Samson— all of which have in common the fact that they represent the attitudes of a group that has no political power or organized professional army to protect it from superior force. The condemnation of these narratives on ethical grounds is an easy and cheap self-congratulation on the part of those who have never known this sort of existence. On the other hand, there is no evidence that this sort of deceitful strategy was ever a normative element in Yahwistic morality—quite the contrary.

73. Compare the covenant with Laban, with Abimelech (Abraham and Isaac), and David's (covenant) relationship with the king of Gath.

74. See below.

defection (with good reason!) from Saul. Later such groups were called *zārîm*, in the Hellenistic period, *lēstai* "robbers" (= *ḥabbāṭū*) and *kakoûrgoi* "evil-doers" (= *pōꜥalê ʾāwen*)![75] Political dissidents with no ethical standards operative outside the in-group were called "pirates." In Elizabethan times groups like the much earlier Robin Hood's band were legally "wolves" = *kalbu ḥalqu*. Such men could be killed with impunity on sight, unless they registered (for a fee) as "king's outlaw."[76] Political states have always produced outlaws,[77] but the ethical value system of the "outlaws" has rarely been of a higher order than that of the state. The Bible is the best illustration of one ꜥApiru ethic which has always been thrown out when the political state took over.

THE ETYMOLOGY OF ꜥAPIRU

In view of the facts that (1) the earliest occurrences of the word ꜥApiru are found by the time of the UR III texts, late third millennium B.C.[78] and nearly contemporary Old Assyrian texts; (2) the word is almost certainly West Semitic and, therefore, a loan word in the earliest Semitic texts of Mesopotamia; (3) the Sumerian word is probably a loan word from Accadian *šaggāšu*,[79] which has little in common with any conceivable etymon for

75. Compare also Prov. 1:10–19, which is a description of a robber gang, though with no discernible political overtones. For the *kakoûrgoi*, see *PRE*, 10:1529; included in the term are: highway robbers, thieves, kidnappers, housebreakers, cutpurses, and, somewhat later, temple robbers and traitors. The *pōꜥalê ʾāwen* fall into the same category, though with more emphasis upon conspiracy, especially in the Psalms.

76. For a good description of the phenomenon see Frank Aydelotte, *Elizabethan Rogues and Vagabonds* (Oxford, 1913).

77. The Muslim Kharijites (those who go out) exhibit some striking similarities, including becoming a separatist state (*The Encyclopedia of Islam*, ed. H. A. R. Gibb et al. [Leiden, 1960–67], s.v.). Compare the *thugga* in India. Finally, for the difficulty of arriving at a precise definition of a robber gang, see the 1939 U.S. Supreme Court decision in *Zanzetta et al. v. State*: "There immediately arises the doubt whether actual or putative association is meant. If actual membership is required, that status must be established as a fact, and the word 'known' [i.e., to be a member] would be without significance. If reputed membership is enough, there is uncertainty whether that reputation must be general or extend only to some persons. And the statute fails to indicate what constitutes membership or how one may join a 'gang.'" W. F. Dodd, ed., *1940 Cumulative Supplement to Dodd's Cases on Constitutional Law*, 2d ed. (St. Paul, Minn., 1940), p. 206. If legislatures and juries cannot define a modern "gang," it is unreasonable to expect a modern scholar to furnish a precise definition of its ancient counterpart on the basis of ancient writings which had as their primary purpose the gaining of political support and military intervention.

78. Greenberg, *The Ḥab/piru*, pp. 15ff.

79. *Ibid.*, p. 88.

ʿApiru, which is the usual Northwest Semitic reading, it may reasonably be concluded that the word's origin goes back to a prehistoric period in Northwest Semitic. Speculation about its etymology is therefore not likely to be a worthwhile occupation. On the other hand, popular etymologies are not only likely, they may even approach proof. The eponymous ancestor ʿEber is certainly an auspicious starting point. When one sees, on the one hand, that the Hebrew word ʿeber means a "party" in the legal (or shall we say quasilegal?) sense in 1 Samuel 14:40, and, on the other, that the personal name ʿEber is assigned two sons, one of whom is Peleg meaning "division" (Gen. 10:25)—the other, Yoqṭān, alone is assigned descendants who are largely identifiable with South Arabic tribes—it seems evident that here is a case of popular etymology. This is further borne out by two facts—first, that a derivative of PLG had already occurred in the Song of Deborah (Judg. 5:15, 16) as a subdivision of the tribe of Reuben, and, second, that Peleg does not have any association with the types of onomastic traditions of the second millennium B.C. as far as we know, but in the genealogy of Genesis 11:16–30 every name occurs in our sources, either as a place or personal name, beginning with Reu, *the son of Peleg*. To reconstruct the process, we might proceed as follows. In the tenth to ninth centuries, to which the tradition as we have it may be assigned,[80] the following facts were known:

1. The term ʿpr was a designation applied to various groups in the old traditions, the shift from p/ to b/ having already taken place in the dialect of the writer.
2. Those ʿpr groups were in part closely associated with the origins of the Israelites; part were not but became the nations of Edom, Moab, and Ammon.
3. Genealogies and tribal relations from remote periods were still current in Israelite tradition.
4. The tradition regarded both the ʿbr groups and the tribal groups as closely related.

The process of reasoning may have been as follows: since the tribes are related to the ʿbr, but the tradition does not trace them back further than Yoqṭān, the ʿIbrî must be the origin, and the name ʿEber the ancestor. A variant tradition regarded the same groups as descendants of Shem, duly recorded in Genesis 10:21. The line which led to Israel, Moab, Ammon, and Edom is a section, division (= Peleg), which corresponds to the fact that Abraham separated from the others in Aram Naharaim. Later, the P source inserted other surviving traditions connecting Peleg with Abraham. The

80. FSAC, pp. 236–43, 251–52.

curious omission of Israel itself in the table of nations can perhaps be explained on the ground that everyone would have recognized in ʿEber the Israelite ancestor. Therefore, may we not conclude that the explanation of ʿIbrî as derived from ʿEber derives actually from the Hebrew word for party, followed by a subsection, Peleg, corresponding to the Hebrew pᵉluggâ? The gentilic is, then, without any *legitimate* etymon in Hebrew, and the astounding similarity in religious, legal, and political status (or lack of it) over centuries between the Israelites and proto-Israelites on the one hand, and the ʿApiru as we have here described the term, would certainly place the burden of proof on those who wish to separate them.

In view, however, of the accumulating evidence for the continuity of West Semitic throughout the Bronze Age, especially from the Byblos syllabic texts, a decipherment and treatment of which I hope to publish before long, it is now possible to give a sound and historically grounded etymon for the term ʿApiru. It is simply the common Semitic root ʿBR, 'to cross [especially a boundary],' the semantic equivalent of the Akkadian etēqu. To cross a boundary illegitimately is to transgress, or aggress. Compare the Latin *transgredior* 'to step across a boundary.' The semantic shift from a political boundary to an ethical/moral one yields the meaning of the term in the Amarna letters. It is preserved in a number of uses in biblical Hebrew, especially ʿābar bᵉrît, 'to transgress a covenant,' but also narratives in Numbers 22:18: "to transgress the mouth [= command] of Yahweh"; and Judges 9:18, which can only mean "to take over in Shechem by illegitimate force," since it was followed by a rite in the temple of the god, culminating in the ritual cursing of Abimelech who at the time presumably was the legitimate ruler. The same process is described in the Jephthah story (Judg. 11:29), where the first uses of the root refer to the act whereby the Israelite areas or peoples of Gilead, Manasseh, and Mizpah of Gilead are incorporated into the movement of resistance (all three are causatives, probably denominatives: "to cause to become ʿApiru"), and the last usage is the act of rebellion against Ammon. Note also the archaic (probably) survival in Exodus 32:27: "to cross a boundary [of the gate] for violence and withdraw" (ʿibrū wāšūbū), as a punitive measure for the worship of the golden calf. Note also the causative of the root in 2 Samuel 3:10, where Abner swears to "alienate" the kingdom from the house of Saul to give it to David.

Arabic has preserved a number of archaic uses which derive from the Late Bronze complex: ʿafarnan, 'insolent and audacious in pride and in acts of rebellion or disobedience.' Note also the meaning from popular etymology 'who rolls his adversary in the dust.' Also ʿifrît/n, 'sharp, vigorous, and effective in an affair, exceeding the ordinary bounds therein, with craftiness, or cunning, and wickedness, or malignity.' The same forms apply to the Jinn with similar meaning. The uses of the later form ʿabar are semantically similar to those of later Iron Age West Semitic, having lost the Bronze Age political and socioethical connotations; compare especially ᶜābir, 'a transient,' 'wayfarer' (Lane, *Arabic-English Lexicon*, *s.v.*).

Finally, compare the semantic development of Babylonian *etēqu*—*itiqtum*, 'transgression' in a political sense, now available in Ugaritica V, and discussed above in chap. I. For the shift in vocalization from ᶜapir to ᶜibr- which has sometimes been used as an argument against identifying the two words, it is only necessary now to cite the identical Canaanite shift (already pointed out by various scholars) from *malik* to *milk* 'king.'

VI

THE "SEA-PEOPLES"

IN PALESTINE

The term "sea-peoples" has long been known to be inaccurate. Originating from the ancient Egyptian texts which describe and name the various tribes of the confederation which swept down from Anatolia and northern Syria, the term has become conventional in scholarly literature. The Egyptian sources themselves refer to families moving by oxcart as well as those military expeditions moving by ship that attempted repeatedly to move into Egypt and take control. It is generally conceded that related groups settled in Palestine and became known as Philistines, a name derived from one of the "tribes" listed in the Egyptian sources, which was also probably the name of a goddess: *mēter plastēnē* of a Greek inscription from Western Anatolia. Others, in the course of time, moved west, where they gave their names to Sicily, Sardinia, and the Tyrrhenians—the Etruscans.

It is here argued that, as usual in ancient or modern history, that which comes to public attention is like the 10 percent of an iceberg which appears above the surface. The magnitude of the catastrophe which struck all civilizations around the perimeter of the eastern Mediterranean induced migration on a scale which we cannot as yet assess adequately. At the beginnings of biblical history we find traditions concerning seven "nations," only two of which have been generally conceded to be resident in Palestine at the time of the Amarna letters: Amorites, associated with the migrations from northeast Syria in the time of Abraham; and Canaanites, who seem to represent the indigenous population of the eastern Mediterranean coastal area from the dawn of history. Four of the seven nations are now most probably or certainly of Anatolian origin. The seventh, the Jebusites, has the closest associations with old Amorite onomastics.

Further study of these populations yields the historical foundations for the peculiar biblical tradition that all mankind springs from a common origin and is therefore subject to a single God.[1] Even though those traditions as we

---

1. The importance of Adam as the "common ancestor" of all mankind is underlined by the universal tendency to give expression to an existing social unity by deriving its members from such a source. See chap. VII.

now have them in the Bible are not very early, they are certainly based upon very old sources and were not simply invented in the Exilic period.

We have learned enough about ancient history to know that the biblical traditions must be taken seriously if not literally. We have an abundance of documentation from excavations comprised of primary, contemporary, and datable sources that yield all kinds of information about the social, political, ethnic, and religious traits of man in the Near East during the period under consideration. Unless biblical history is to be relegated to the domain of unreality and myth, the biblical and the archeological must be correlated.[2] Methodologically, the archaeological documents, especially the written ones, must be given priority and considered seriously. Far too many correlations between biblical and archaeological data already exist to justify any a priori concept that the beginnings of Israel in the time of Moses and Joshua constitute a *tabula rasa*, a complete discontinuity from that which preceded it. Just as Greece constitutes a continuity of the Minoan, Mycenean, and Anatolian cultural traditions, and Assyria a continuation of Amorite, Hurrian, Sumerian, and Babylonian traditions, so also Israel represents the creative transformation of pre-Israelite cultures, a new synthesis dominated and inspired by a new value system: a complex of convictions which in some measure determined human behavior. What those convictions were can best be described within the framework of historical narrative. In biblical thought, the value is inseparable from and can be known only through its expression in action—only in this way does the value become historically real. If the value is a religiously binding obligation, the fact that it does not become historically real, i.e., observable and recognizable by a third party, i.e., God, means then that the consequences of such neglect become historically real: the divinely imposed curses. Not until the elaboration of liturgy does a third alternative become too easily possible—and in this respect the prophetic denunciation of ritual represents the continuity of the earliest nonpagan traditions of the Bible.

Even in modern times, it is clear that inclusion of a large population group under a single, unified political regime does not necessarily make them "neighbors," i.e., neighborly. In ancient history this seems to have been even more true, for identification of persons and social groups based upon political citizenship seems to be rarely attested,[3] except by outsiders,

2. Contrast James Barr's *Comparative Philology and the Text of the Old Testament* (Oxford, 1968). His whole argument seems to constitute a systematic protest against using any but traditional methods in interpretation of the Old Testament. In 1951, Pfeiffer could still say in a presidential address to the Society of Biblical Literature, "Not only scholars, but even the humble untutored believers of all faiths intuitively know that facts and faith do not mix" ("Facts and Faith in Biblical History").
3. Mendenhall, "Relation of the Individual to Political Society," p. 97.

and not long-lived. On the other hand, all city-based political societies of ancient times, particularly in Palestine and Syria, demonstrate the same sort of population diversity characteristic of modern cities all over the world. Though America claims to have invented the "melting pot," most of history constitutes this process,[4] and ancient Israel was no exception. Racism is a most thoroughly modern invention, and we should know, to paraphrase the old legal maxim *summum jus, summa injuria*, that "the purer the race, the purer the evil." Which "race" we are talking about is not of the slightest significance. "Racial purity" is a concept of value only to the racist.

The Late Bronze Age saw the transition in Palestine from the seven nations to the twelve tribes; in the coastlands, from the Canaanites to the Phoenicians and the Philistines; in the inland area, from a complex of Amorites, Hittites, and Aḫlamu to the Arameans and the Syro-Hittites. In a bygone era of scholarship, such transitions were explained by the hypothesis of "waves of migration" of some primitive, "pure," fresh, and unspoiled tribes out of some land of Oz which somehow never could be located. It is undeniable that migrations did take place before, during, and after the period with which we are dealing—such migrations are a constant in all human history, with fluctuations in intensity at different times. Yet, in spite of the evidence for migrations, it is hardly possible to work now with any other hypothesis than that we are dealing with periods of such rapid social and cultural change that researchers were unable to recognize the fact, preferring instead to resort to racist ideologies to explain the observable contrasts.

THE "SEVEN NATIONS"

Biblical tradition holds that prior to the Conquest, Palestine was divided among seven nations, to use the term traditional in biblical scholarship. It is immediately apparent, however, that a great diversity exists among the twenty-six passages enumerating the various "nations" which inhabited Palestine before the formation of Israel. The number ranges from one to ten, the order exhibits great variation (only two clusters of six "nations," with three examples of each; in both cases the three are assigned by literary critics to three different "sources"),[5] and the nations listed vary also. It is impossible to conclude that the list is based upon anything other than historical fact—

4. One need only think of the great diversity of population and language groups that made up the people of the British Isles.

5. Deut. 20:17 (D); Josh. 9:1 (J); Josh. 12:8 (E?). Cf. Deut. 7:1; Exod. 3:8 (J = 3:17); Judg. 3:5 (D?); Exod. 13:5 (L). There is no evidence for a "canonical list," and of twenty-six different lists of the pre-Israelite population of Palestine, only three lists in fact mention seven nations: Josh. 3:10 (J); Josh. 24:11 (E); and Deut. 7:1 (D). (The "complete" list is found twelve times in the LXX.) These three agree, so far as the

which was itself changing and was seen in different ways at different times and places within ancient Israel. In other words, it was not a mere "canonical tradition" with no historical basis. The historical problem is to correlate the various terms with some kind of sociopolitical reality which underlay and gave rise to the biblical tradition: what were those one to ten "nations" and who were they that they could largely disappear during the course of the thirteenth–twelfth centuries B.C.?

The Canaanite problem seems to have been solved by their identification with the Phoenicians decades ago; similarly, the Amorite problem was solved with the discovery of the Mari texts—and yet the biblical usage does not consist of data which will enable us to demonstrate any contrast between the two peoples. The Jebusite is identified, so far, only with the pre-Davidic population of Jerusalem, and can be correlated also with an Amorite name from Mari, *Yabusum*; it should also be correlated with the Transjordanian name, Jabeš-(Gilead), which equals Amorite *Yabišum*. The two were dialect variants in West Semitic by the time of the Middle Bronze Age, ca. 1700 B.C.[6] What the name meant we do not know.[7]

The remaining four nations have not been correlated successfully with any known Semitic population group. The Hittites of course are well known in Anatolia in the Late Bronze Age, but it has been generally and wrongly held that the population of the Hebron region had little or nothing to do with the historical Hittites.[8] The Girgashites have been found in both Ugaritic and Egyptian sources of the Late Bronze Age, and constitute therefore a historically attested social group of some sort which we cannot further define or locate. There remain, then, the Hivvites and the Perizzites, either unidentified or abolished by textual emendation.[9] Leaving the Perizzites aside for the moment, it should be observed that the Hivvites are quite thoroughly and concretely attested in the biblical tradition. They are assigned to specific localities, which is unfortunately rare in the case of most other nations, namely, to Shechem, Gibeon, and the Biqᶜa of Lebanon.[10]

---

order is concerned, only in placing Jerusalem last (Jebusite). The conclusion is inescapable that these three lists stem from the period of the United Monarchy.

6. *Ya-bu-sú-um*, *ya-bu-si-um* (*APN*, p. 38); *Ya-bi-šum* (*OK*, p. 24); and cf. *APN*, p. 178, where the latter is rather tentatively related to the root BŠ. See also *APN*, p. 177.

7. In view of the well-attested occurrences in Old Amorite, a non-Semitic explanation seems improbable, though the presence of both *Iabus-* and *Peraz-* among the twelve "tribes" of the old Umbrian town of Iguvium in Italy (3rd century B.C.) may well be more than a wildly improbable coincidence. See J. W. Poultney, *The Bronze Tables of Iguvium* (Baltimore, 1959), pp. 190–92.

8. *IDB*, s.v.

9. So Speiser, in *IDB*, on the Hivvites, s.v.

10. Gen. 34:2; Josh. 9:7 (*Xorraios* in the LXX); 11:19 (the phrase, "except the Hivvites, the inhabitants of Gibeon," is not LXX); Judg. 3:3.

Except for the Jebusites, they are more concretely attested than any other of the seven or ten nations, yet it is improbable that they can be identified with any known Semitic society.

There remain quite a number of other possibilities which should be examined since we know of migrations of peoples from the north both in Hittite and Egyptian sources.[11] We are dealing of course with the problem of the "sea-peoples," to use the Egyptian term that correlates well with the Greek tradition of the Pelasgians. Both are very shadowy, obscure "ethnic" groups, but they seem so only because of the modern obsession to read racial, ethnic identity into virtually every social term used in ancient languages. We would have equal difficulty in determining their identity if we suffered under the delusion that the term "pirate" designated some pure racial stock. Even in modern times, the tradition that imperialists always come by ship, not overland, persists. It is a most tempting solution to the problem that the term sea-people arose at a time when armed bands transported by ship were roaming all over the Mediterranean from various home bases, not merely for purposes of piracy, but for the specific intention of relocating and assuming military control of a subject population. At the same time, other groups proceeded overland, and we have then the curious fact that the "sea-peoples" invading Egypt come by wagon as well as ship.[12] They are classified together because the Egyptians knew them to be related population groups, though the relationship need be nothing but geographical. That they are a "confederation" implies a unification by covenant.

It remains to be seen whether or not the term arose at a time when new military techniques or armaments made possible extensive and successful naval landing operations, or whether the hitherto peaceful sea trade had disintegrated and looting had taken the place of trade. Or were population pressures, famine, and disease responsible for a widespread desire to emigrate? All of these factors could probably be illustrated from the narratives of Herodotus about much later migrations, and probably all played a part. It is not possible to generalize about the cause, but concerning the process involved it seems quite justified to give a hypothetical "ideal model" in somewhat the following fashion: (1) Seafaring adventurers engage in trade and occasional looting in far-flung places. (2) Development of superior armaments and military techniques makes it possible for a very small, well-organized

11. For Egyptian sources, see *ANET*, pp. 262–63. For a Hittite source, see E. von Schuler, *Die Kaškäer* (Berlin, 1965), p. 36 (as slaves?), p. 38: Kuruštama people had settled in Egyptian-held territory.

12. H. H. Nelson, "The Naval Battle Pictured at Medinet Habu," *JNES* 2 (1943): 40, 43. Nelson's conclusion that the Egyptians knew them to be a coalition of various people, not a homogeneous body (*ibid.*, pp. 44–45), needs to be strongly underlined. Cf. also Schaeffer's discussion in *Ugaritica*, 5: 638–70.

fighting force to subjugate numerically superior groups. (Cf. iron weapons at the Late Bronze–Early Iron transition, and the composite bow and chariot at the Middle–Late Bronze transition periods.) Further, the local village groups are usually devoid of either the necessary skills or weapons for serious warfare. (3) For various reasons permanent migration from the home base seems desirable or advisable (including political reasons—cf. the great concern with fugitives in Late Bronze Age treaties). (4) The armed band moves in by ship or land and establishes control over an essentially non-military village and shepherd population. (5) It becomes the political and military elite, what might be called a superstratum, corresponding to the frequently used concept of substratum, but in a different context. (6) Often cut off from their original homeland, they intermarry with the local population.[13] This eventually results in the fusion of the foreign with the native population, and before long it is only certain formal aspects of the total culture which can be traced to the foreign origin, especially proper names and political designations, for the new regime of number 5 is usually regarded as the continuation of that of the homeland of the superstratum. Its "god"— often the "genius" or *tyche* of the home city—is identified with that of the native population, and the result is what we term syncretism.

That other factors may well have played a part in the complex process there can be no doubt, e.g., employment of armed groups as mercenaries in a foreign land, but the process is both illustrated in detail from texts of various periods, and, so far as the population fusion is concerned, illustrated by foreign occupation armies all over the world in spite of modern racism. From this perspective, the seduction of Dinah makes good historical sense— as does also the indigenous reaction. The point of the narrative, however, is that the newly arrived Shechemites *wanted* to regularize the intermarriage by covenant relations. A thousand and more years later the veterans of Alexander the Great's empire found themselves in the same sort of situation after his death—cut off from their homeland and wives both by enormous distances and political hostilities—and mass marriages resulted.

In view of the fact that Rib-Addi of Byblos pleaded desperately for a platoon of forty Egyptian soldiers to protect his city from the hostility which was threatening to overrun it and separate it from the control of pharaoh,[14] it would have been equally possible for one or two shiploads of well-armed warriors to move in and take over the small city-states of the coastal area.

13. This is the context of Gen. 34. Compare the mass marriage between the Greeks and Persians arranged by Alexander, according to Arrian, *Anabasis* vii, 4.

14. *EA* 108:66. This seems to be a minimum. Elsewhere he pleads for 30–50 teams of horses plus 200–500 troops (*EA* 71:23–24; 85:19–20; 127:36–37). Further south, the usual request for garrison troops consisted of a squad of 50 men (*EA* 238:11, 289:42 [Jerusalem]; 295:r 6).

There is no reason to believe that the same was not true of inland areas, considering that most of northern Syria as far south as Hama became politically Syro-Hittite after the collapse of the Hittite Empire in the second half of the thirteenth century B.C. This period coincides very well with the collapse of the Mycenean Empire in the Aegean region, and with the traditional (in biblical scholarship at least) date of the Israelite "Conquest."

It has already been pointed out and long since accepted, that the Mitanni state was largely dominated by a thin superstratum of Indo-European aristocracy, with a population of Hurrian- and Semitic-speaking peoples.[15] The same is true of Palestine and southern Syria in the Late Bronze Age, though with a much smaller proportion of Hurrians than further north, nearer their main population center in eastern Anatolia. Cilicia, according to Goetze, received an influx of Hurrians after 1500, and after 1200 was dominated by Hurrians.[16] The proper names from both Ugarit and Alalakh show Indo-European, Hurrian, Semitic (both Amorite and Canaanite, though by the Late Bronze Age we frequently cannot distinguish between these name groups), and Luwian origins.[17] Though we have very extensive and detailed information from northern Syria and southern and central Anatolia dating to about 1250 B.C. or shortly afterward, we have next to nothing on which to base conclusions, for Palestine and Transjordan, from contemporary written documents after the Amarna age, ca. 1360 B.C. Yet it is not until Merneptaḥ, ca. 1225 B.C., that we hear of the first conflict with the sea-peoples in Egypt,[18] and not until a half-century later that Ramses III engaged in a desperate sea and land battle with such invaders.[19] Though by the first half of the fourteenth century B.C. there were already population groups in Palestinian cities deriving from the north and in control, the evidence from Egypt strongly suggests that this was a trickle compared with the flood which came a century later.

Whatever may have been the historical dynamic which had the same effect in Greece (virtually every excavated site was destroyed by violence between 1250 and 1150 B.C.),[20] Anatolia, and Palestine, one can no longer hold that the process of history in one area is hermetically sealed off from

15. R. T. O'Callaghan, *Aram Naharaim* (Rome, 1948), pp. 51–74.

16. A. Goetze, "Cilicians," *JCS* 16 (1962): 48, 54.

17. *PTU*, 268–97. Hittite and other Anatolian names are of course included in this section. It goes without saying that the list is minimal: there are certainly a considerable number of Anatolian names that have not been identified.

18. Note that Sherden were already serving in the Egyptian armies by the time of Ramses II (*ANET*, p. 255; cf. Gardiner, *Onomastica*, 1: 194*). They usually seem to have been captives taken in war who were subsequently integrated into the Egyptian army.

19. See n. 12.

20. Webster, *From Mycenae to Homer*, pp. 136–37 and p. 325n.

that in another (except by the peculiar accidents of modern academic departmental organization). Both Near Eastern and Greek traditions insist upon widespread migrations from Anatolia at a time which is, from the point of view of the various writings involved, prehistoric. This period can now be seen in archaeological perspective to have been merely the end of another archaeological and political era—the end of the "First International Period" as it has been known since the days of James Henry Breasted.

The destruction of empires and the consequent disruption of social order, economic organization (especially of the cities), and political and even personal security meant that many persons had to find refuge elsewhere. The military adept would move to some new area and become the military aristocracy. Most merely squatted on territory which furnished some potential for economic security at least, and such territory was to be found in the desert fringe which extends from the Hejaz of the Arabian peninsula to the Euphrates River, through Transjordan. It is this area which seems to exhibit a dramatic increase in population at the transition period from the Late Bronze to the Early Iron Ages,[21] though much more investigation is badly needed. The Arabian Desert is most emphatically not the reservoir or source of successive waves of migration (as maintained by traditional nineteenth-century concepts). Rather, the desert fringe was the frontier for those whose security from the man-made catastrophes of history could be obtained only by flight. An additional impetus to such dispersion must have been the outbreak of epidemic disease, for virtually every one of our collections of sources from the Late Bronze Age refer to such epidemics as very grave concerns.[22]

If, then, the Amarna age documents demonstrate for us the presence of politicomilitary superstrata which originate in the north, the historical situation a century later should be characterized by a much greater domination of small population groups of northern origin. Transjordan in the Late Bronze Age had very little population, though it had great economic potential. In the early phases of the Iron Age, its population must have increased many times, and the orientation of the culture is toward the north as well as the west, so far as evidence is available.[23] But archaeological artifacts have no necessary correlation with geographical origin of populations, and even less with ethnolinguistic identity. One shudders to think of trying to identify the ethnic identity of the population of Ringold County, Iowa, from archaeological materials of, say, A.D. 1840. Ethnic identity is a function

21. *EEP IV*, vols. 25–28, pt. I, p. 423. However, it has been demonstrated that the conclusions in this work are not all reliable.

22. See chap. IV.

23. J. B. Pritchard, "New Evidence on the Role of the Sea Peoples...," in *The Role of the Phoenicians in the Interaction of Mediterranean Civilizations*, ed. W. A. Ward (Beirut, 1968).

of personal relationships and interests, combined with such similarities of language, culture, and religious value systems as well as geographical origins, as make new functional social organizations possible. Rural Iowa of 1840 is remarkably similar to Palestine in the thirteenth–twelfth centuries B.C. Even their lamps (called "sluts") have their closest analogy in the Late Bronze–Early Iron Age lamps—a strip of cloth hanging over the edge of a shallow dish filled with some available kind of inflammable fat. The difference between the two eras consists very largely in the fact that there was no very great cultural disparity between the old and the new populations in Iron Age Palestine. In the Late Bronze Age it is impossible to conceive of a population anywhere in the fertile crescent which had not for generations been in intimate contact with the city-states of that period. The barren desert of Arabia has not yet yielded any evidence of occupation of any sort until the late Bronze Age, though this is very probably the result of inadequate exploration. Since the camel is not indigenous to the Near East, the camel nomadism which introduced for the first time a true nomadic culture to the desert fringe is almost certainly a non-Semitic, extraneous culture trait, introduced via Anatolia from somewhere further to the northeast. The nomadic mirage must finally be laid to rest. Nomadism has no more to do with the origins of the Semites or of Israel than it has to do with the origins of Greece and Rome; indeed, there is probably less ground for such theories, in view of the fact that much more evidence exists for Indo-European nomadism than for the Bronze Age Near East, though admittedly of a later date. Animals were important to all cultures of the Bronze and even earlier ages, as well as much later. But the mere possession of animals does not constitute a nomadic society in Bronze Age Palestine any more than it does in Caesar's Rome.

Growing archaeological evidence points to a rapid increase in the population density of Transjordan, especially during the Early Iron Age, and the same is probably true of Palestine (though much more sophisticated archaeological and demographic investigation needs to be done before secure conclusions can be reached). After all, the empires of the Late Bronze Age were very complex political, economic, and social (as well as ideological) structures, and an event recovered in a destruction level in one particular site cannot be interpreted historically apart from that total structure.[24] Nor can changes in artifacts alone be used as evidence for population changes, except under the monstrous hypothesis that the ancient peoples involved were absolutely incapable of any kind of economic, technological, or social change.[25] Evidence from sudden changes in architectural patterns are some-

24. Note the repeated requests for Egyptian garrison troops cited above in n. 14.
25. Cf. A. Goetze's statement that it would be impossible to demonstrate the presence of an Assyrian trading colony at Kaneš/Kültepe on archaeological grounds alone (*Kleinasien* [Munich, 1957], pp. 74, 80).

what more persuasive, but diffusion of fashions in this area, apart from population change, could also account for discontinuities. A rather rapid deterioration of quality in artifacts such as pottery (from the modern technological point of view) proves absolutely nothing concerning ethnic identity of potters or consumers. It demonstrates merely that the demand for quality (again technologically) products in the domain of pottery-making no longer existed. If changes occur in the shapes and techniques of pottery, it indicates nothing more than that the professional specialist quit in disgust or abandoned his craft quality standards in order to remain in business at all. If techniques changed, it is perhaps because the very poor neighbor next door who did not know how properly to make pottery suddenly found a new source of income, or even more probably because women took over the job when the men were busy at war. Far-reaching conclusions about ethnic migrations drawn on the basis of such artifacts *alone* are completely inadmissible, but when such changes can be correlated with pottery forms *and* techniques which are incontestably identical to those from some distant area—and if it can be proven that they are not imports—then the evidence may be used (if it correlates with other kinds of evidence—archaeological, historical or linguistic).

We are in an increasingly fortunate situation in this regard, because of the rapidly growing mass of historical and artistic statements from all over the eastern Mediterranean, plus the archaeological evidence of destruction in Palestine itself. The disintegration and almost universal destruction of civilization in the entire area, accompanied if not caused by mass migrations— probably radiating in all directions away from the epicenter of the cultural shock—meant that new forms of social organization were inevitable, and were actually developed with great rapidity. After all, even in the slow- moving social organizations of the Orient, it would not take much more than a few days for groups to establish some sort of covenant relationship which might become functional enough to survive for some time. Alternatively, an armed band could "pacify" a region and establish its own petty chiefdom in a few hours of confrontation in a city-state area, which is very probably how the kingdoms of Sihon and Og were established. The result is then an inevitable blending of what were originally diverse cultural and linguistic traditions, but cultural diffusion for a thousand years prior to the catastrophe makes it impossible to conceive that there was nothing in common between the two groups.

In the first place, tribal and political boundary lines were completely irrelevant so far as the diffusion of the cult of ancient gods is concerned. In the Bronze Age, the cult of Hadad, Dagon, and ꜥAnat had already spread from Mesopotamia to Egypt, with translated counterparts in the storm gods of the Hurrians and the Luwians, to say nothing of Zeus who is Greek in name only. In addition, the function of the gods in overseeing the sanctions of covenant relationships was also a common factor in civilizations all over the eastern Mediterranean area, as the Hittite and Egyptian treaties demonstrate beyond question. It is this that furnished at least the possibility of recognizing a transcendent factor in human experience and culture. The gods could not be made merely symbols of the existing concepts of the political and economic interests of the state with its primitive tribal mentality. It must be admitted, however, that in the Late Bronze Age this transcendent factor was hardly more than a potential, which is so well illustrated in the fact that the same god was localized in many states or societies but was not necessarily functional in the relations between them. The various Hadads, Ishtars, sun-goddesses, and so on are distinguished by their cult-centers in our texts. If their major function was merely as symbols of existing interests, the identity of name had no further social or historical significance. The problem was faced also by pre-Exilic Israel, for Deuteronomy had to emphasize very strongly that there is only one Yahweh. (Not until the *destruction* of state and temple was there a possibility of rediscovering the Mosaic oneness of God, and the recognition that God is more than merely a political symbol of unity.)

The second factor which enormously facilitated a new religious and cultural synthesis after the catastrophe was the evidently complete insensitivity of ancient man to the sorts of racial or ethnolinguistic contrasts which so plague the modern world. Indeed, it is difficult to see how such contrasts (other than linguistic ones) could have been systematized in view of the universal diversity of language which all our Late Bronze Age documents prove. The analysis and recovery of such linguistic evidence, which alone would be decisive in proving large-scale migration, is in its infancy but has made extremely rapid progress in the past two decades. The application of the evidence to the biblical materials has hardly begun, and the few observations published here constitute not much more than a vigorous suggestion that the field has a very considerable backlog of unfinished business.

As is argued further in the next chapter, scholarship must begin to make a distinction between political organizations, whether of the tribal or state nature, and ethnic groups or races. Particularly in the case of the sea-peoples, the idea that they constitute some pure racial stock which contrasts systematically to all others is outside the realm of possibility. If the Philistines are termed "the uncircumcised" in the Bible, the Egyptian representations prove that at least some of the sea-peoples were circumcised. If circumcision is an evidently obligatory marker of the group (not a covenant obligation originally), the narrative of Joshua 5:2-9 proves that at least a considerable segment of what constituted the religious community at that time was *un*circumcised, as the very existence of the narrative and the elaborate explanation proves. To later generations it was inconceivable that they could have been uncircumcised, but the linguistic evidence suggests strongly that their origin was just as much from the north as was that of the "Philistines." The contrast between Philistines and Israelites was therefore purely political, religious, and cultural, not ethnic or linguistic. Both were new social organizations of fairly large scale, but with radically differing value systems. Both were new organizations of the *existing* population groups, for the small band of migrants from Egypt had become absorbed or integrated into the Transjordanian populations before the death of Moses, just as military adventurers from the north had become integrated into the political and social structure of Palestine and Syria by the time of the Amarna Letters. Those who migrated south did not bring with them a complete social, economic, and political organization; they were by no means homogeneous in their lands of origin, and they were not a unified group which acted in total concert against existing populations. There were Semites and sea-peoples on both sides of warring armies, just as both had long been warring among themselves in their respective places of origin prior to the catastrophe, whether in the land of the Hittites or Aram Naharaim.

After all, the major purpose of the old pagan states and empires was to *include* as much as possible under their control for political power and economic exploitation, and primitive tribal exclusiveness was eliminated by the Yahwistic movement, if it ever existed on any important scale in the Bronze Age. Tribal organizations must have existed (of the political type described in the next chapter), and they probably practiced endogamy. The Yahwistic revolution, however, created a situation which made possible intertribal marriage from Dan to Beersheba. The only prohibition of marriage outside the group was aimed specifically against the old pagan custom of using marriage as a bond between two groups for the purpose of political power ploys—so well illustrated in the policies of Solomon and some of his successors, as well as in the ridiculous political maneuvers of Abimelech which led to catastrophe, not only for himself but for the whole city of Shechem. Such marriages are simultaneously a subjection of marital relationships to political ends, and a sacrifice of ethical obligations under the covenant to political goals, as pointed out in Exodus 34:12–16 and Deuteronomy 7:3–5. Both passages append the warning against intermarriage with the various "*goyim*" to the primary warning against making international treaties with them. The earlier passage in particular is a summary of the events at Baal Peor which had such disastrous consequences. The refusal of social groups to make covenants with more powerful military regimes is more or less predictable and is attested elsewhere in the ancient world.

THE HIVVITES

The Hivvites have been an enigma since antiquity, but unlike some of the other social groups mentioned in the Bible, the accidents of source discoveries now make it possible to suggest an identification which fits in with everything we know and actually clears up a number of rather peculiar facets of biblical history. They are included in every list of the six, seven, or ten nations except in Genesis 15:21 and Neh. 9:8, and therefore it is very difficult to abolish them by textual emendation, as the late E. A. Speiser seemed strongly inclined to do in the *Interpreter's Dictionary of the Bible (s.v.)*, on the ground that *Xorraios* occurs in the Septuagint version of Joshua 9:7. In view of the fact that the Hivvites are mentioned three times as often as the Horites, and in view of the fact that the latter are attested only for southern Transjordan (outside of the Septuagint Josh. 9:7) it would be much more probable that the Horites can be explained away, if necessary, than the Hivvites. In fact, neither has to be expunged from the biblical record, but both must be fitted into the total context of the history of Palestine and Transjordan so far as can now be determined.

When Winckler prepared the third edition of Schrader's *Die Keilinschriften und das Alte Testament* (1903), the identification of biblical Quwe (Kue) with Cilicia (1 Kings 10:28) was already known.[26] It was not until decades later that Albright demonstrated that the cuneiform *ḫu-we* is identical to *ḫu-me*, citing at the same time a number of examples of a shift from *q/k* to cuneiform *ḫ*, and shortly afterward, the Babylonian Chronicle confirmed this identification.[27] Since the shift from **ḫu-wi* to *ḫi-wwi* is negligible in view of numerous shifts from /u/ to /i/ in West Semitic dialects (and vice versa), the identification of the Hivvites of the Bible with Quwe/Cilicia must at least be seriously examined.

It is a curious and inexplicable fact that the Hivvites are never mentioned in sources which the literary critics assign to the P document of the Pentateuch, though J, E, D, and Eissfeldt's L documents all mention them repeatedly. Curiously, in every enumeration assigned to J, the Jebusites constitute the last one of the series, but of the passages assigned to E, only in Joshua 24:11 do they find their place at the end of the list, presumably of course the climax. As one might expect, the Deuteronomist likewise places the Jebusites last in both cases where he cites earlier sources (20:17; 7:1); however, compare Judges 3:5, presumably Deuteronomic, where the Jebusites

26. E. Schrader, *Die Keilinschriften und das Alte Testament*, ed. H. Zimmern and H. Winckler, 3rd ed. (Berlin, 1903), p. 238.

27. W. F. Albright, "Cilicia and Babylonia under the Chaldaean Kings," *BASOR* 120 (1950): 22–25; "The Nebuchadnezzar and Neriglissar Chronicles," *BASOR* 143 (1956): 28–33.

also end the list, but the first three occur in quite a different order similar to J and L of Exodus 3:8, 17 and 13:5, respectively. What all this leads to is the conclusion that the traditions about the seven nations do not stem merely from some legendary formula, but are based upon social and political realities of the early period before the monarchy. Since the "nations" had their bases in the various city-states which did not become a part of the Israelite confederation, there would have been a constant historical base for the traditions of the seven nations until they were finally subjugated in the successful conquests of King David. Those traditions may well reflect the changing political scene of the period shortly before David's conquests—not the time of Moses. By the time of the Priestly writer, they had become presumably completely irrelevant, and are thus not even mentioned—or better, he had nothing to add. Probably, even the proper pronunciation of the name had been forgotten.

The constant tendency of biblical scholars to identify every gentilic in ancient sources as a designation of an "ethnic group" based upon race is a most serious handicap to progress in historical reconstruction. Wherever there are sources, the evidence of the fact that this procedure is completely inadmissible is abundant. Gentilics are most characteristically derived from local place names, especially cities and villages.[28] Second, they are derived from the designations of political power systems. Third, they come from what are now called "tribes," but in the absence of the kinds of data by which anthropologists can classify nonpolitical social structures, we cannot go much further than merely to observe that a tribe is a solidarity group, which often regards its members as descendants of a legendary ancestor, functions in certain ways for the interests of the entire group, and constitutes the larger community of which the villages are a part.[29] This kind of social structure is characteristic at early stages of the entire Mediterranean world, and usually furnishes the foundation for the development of the politically organized state (and indeed often furnishes the name for the state as well, for which the best illustrations are the names Judah and Israel during the Divided Monarchy).

The names of the seven nations are accordingly to be regarded as political not ethnic designations. They are the major power centers of the period of conquest, and as we can see in the Amarna letters, such power centers rise and fall as historical factors with considerable celerity. It is not to be expected that the list of "nations" or their order would be identical in all periods prior to the monarchy if the lists are based upon historical reality (as we believe they must be).

28. Mendenhall, "Relation of the Individual to Political Society," p. 93.
29. See chap. VII.

It has already been argued above that we have every reason to believe that an increasing number of peoples from the north came into the south-eastern Mediterranean area in the course of the Late Bronze Age. In fact, it seems impossible to point to any period within the known history of Palestine when we do not have evidence of intimate contacts with Anatolia, whether it be the pre-pottery Neolithic, the Bronze Age, the Iron Age, or the nineteenth-century Ottoman Empire.[30] The migration of peoples from that area cannot therefore be regarded as a mere single episode of ancient history, and no arbitrary beginning point or ending point can reasonably be expected. Datable evidence does however give us some indication of what had recently happened.

Hittite sources earlier than the Amarna age already refer to the migration of people from Anatolia to territory under Egyptian control.[31] It is not surprising, therefore, to find in the Amarna Letters a statement from the king of Jerusalem, Abdu-Hepa (at least a half-Hurrian name from northern Syria), complaining that Milkilu, king of Gezer, does not sever his relationship with the sons of Labayu at Shechem and "with the sons of Arzawa."[32] Arzawa here lacks the personal determinative, and certainly must be taken to be a reference to that important segment of Luwian-speaking peoples who resided somewhere to the northwest of Cilicia. At this time, Kizzuwatna itself must have been fairly evenly divided between Luwian and Hurrian populations. The implication of this passage is that the population of Shechem in the first half of the fourteenth century B.C. had a politically important segment which came from an area not yet precisely defined, but which was intimately related linguistically to the Luwians of Cilicia = Quwe = Hiwwi. Chapter 34 of Genesis calls Shechem (the town name which occurs already in the Egyptian Execration texts of the eighteenth century B.C. is equated with the name of an eponymous ancestor) the son of Hamor,[33] "the Hivvite." Further, archaeological affinities between Shechem and Anatolia have already been pointed out.[34] On the other hand, we have no more specifically Luwian or Arzawan names from Shechem than we have "Philistine" names from

30. Note also Albright's suggestion that there is evidence of the presence of Lycians in Byblos as early as the Middle Bronze Age ("Dunand's New Byblos Volume: A Lycian at the Byblian Court," *BASOR* 155 (1959): 31–34.)

31. Cf. n. 11 above.

32. *EA* 289:5–7.

33. The name is again Anatolian: Ḥamri designates a cult installation of Cilicia; cf. *DLL*, pp. 129–30. Popular etymology accounts in part for the Semitic form, but compare also paḫri = peʿōr (chap. IV), and *Pitru* = *Petōr*; therefore, ḫamri > ḫamor. The word must certainly be ultimately a loan word from Akkadian, *AHW*, s.v. "holy precinct."

34. G. E. Wright, *Shechem: The Biography of a Biblical City* (New York, 1965), p. 97, Alalakh; p. 107, Boghazkale; p. 121, Boghazkale.

the Philistine cities. The initial wave of conquerors had left little but archi-tectural innovations and pseudoethnic names behind by the end of the Iron Age. The list of proper names from prebiblical Shechem exhibits the same kind of linguistic rainbow we find in contemporary Ugarit and in the Amarna Letters:[35] in other words, even if the Shechemites were Luwian in speech and political tradition, there is no ground whatever for thinking that they were a pure ethnic-tribal stock even prior to their migration to Palestine. The same is true of the Moabites, Philistines, and above all of the Israelites, whose movement was a conscientious and religioethical transcending of all tribal-political contrasts until it sacrificed the religious and ethical concerns in favor of a political power structure in the time of Solomon, which was equally intolerable to Israelite and non-Israelite alike by the end of that reign. It was not the first and, tragically, by no means the last time a noble resistance movement degenerated into conformity to the same sort of bureaucratic and power-mad egocentrism which called forth the resistance in the first place.

When we look at other sites in the Palestinian-Syrian coastal and inland areas, we find ample evidence of the same kind of historical process. After the collapse of the Hittite Empire, the traditions and even proper names associ-ated with Anatolia spread far to the west and to the south as well. The northern part of Syria was even called *māt Ḫatti* in Assyrian times, and as far south as Ḥama (even further south to Restan/Rustam) hieroglyphic Luwian writing is attested even as late as the seventh century B.C.[36] Even by the Greco-Roman period, Luwian names persisted as far south as the Damascene, though the possibility cannot be excluded that the name Aldamoas may be specifically a Lycian intrusion which is later than the Early Iron Age.[37] It is not adventuresome at all, then, to suggest that the biblical traditions are entirely correct in attributing to Cilician origin certain dominant population groups in several of the towns of Palestine and the area north of the Sea of Galilee.

The archaeological record is of particular interest in this regard.[38] At the time when northeast Syria was losing or had lost its impressive density of population, Transjordan's population skyrocketed. Though surprisingly few Late Bronze Age sites have been identified there, the Early Iron Age popula-tion was already very impressive.[39] Certainly much more precision must be

35. *Ibid.*, pp. 208–13.
36. Harald Ingholt, *Rapport préliminaire sur . . . fouilles à Hama* (Copenhagen, 1940), 2: 115–16.
37. LPG, pp. 113–14, summarizes the widespread evidence from the Iron Age outside Anatolia.
38. Pritchard, "Role of the Sea Peoples."
39. EEP IV, vols. 25–28, pt. I, p. 423. But compare n. 21 above. Far too little excavation has been done to support this statement, which must be regarded rather as

brought into the discussion of population fluctuations through excavations, but it is not probable that the evidence from surface exploration will be modified other than quantitatively. When we ask, however, where this population came from, the answer is not nearly so simple as biblical fundamentalism would have it. The northern affinities of that population from excavated sites is beyond doubt, and the present writer would strongly suggest that this evidence should be correlated with the Jacob traditions of Genesis. Before Israel existed there had already been migration from the north and northeast, an independence from existing political structures, and integration into population groups already present in the Palestinian and perhaps Transjordanian areas. Of course, the extraordinary incidence of pottery from Aegean and Mycenean areas does not by itself constitute proof of migration, but when this is correlated with massive architectural features such as the Tell es-Saʿidiyeh waterworks (and the Boghazkale parallel), the gold leaf jewelry from the Amman airport and its temple with Iranian affinities,[40] and many proper names later to be discussed, it would be obscurantism of the worst sort to refuse to consider the possibility of important migrations from the north, which included the kinds of populations illustrated so impressively by the name lists of Ugarit and Alalakh. This is particularly true when the Amarna period, a century before the Bronze–Iron transition period, exhibits an impressive number of non-Semitic northern proper names among the superstratum.

The beginnings of the specifically Israelite (i.e., Mosaic) traditions in the Bible show four kingdoms in the Transjordan area. To the south are the Edomites, equated by biblical tradition with the Horites but now usually taken to be Hurrian. Edom is Śēʿir which can easily now be equated with Syrian (Hittite) Šaḫḫiyara, and its biblical explanation meaning "hairy" would fall, together with virtually every other biblical explanation of ancient names, into the category of popular etymology. It is to be expected that foreign names, and also names which are so ancient that their original meaning was forgotten, would be fitted into later linguistic patterns and reinterpreted. It is for this reason, among others, that the clarification of

---

a prediction. The University of Michigan 1971 survey of sites in the Euphrates River valley above Raqqa, Syria, has so far yielded evidence of only three Early Iron Age sites out of some seventy-five investigated.

40. *PEQ* 98 (1966): 155–62. G. R. H. Wright, *ZAW* 79 (1966): 352–57. Campbell and Wright, *BA* 32 (1969): 104–16. The very exotic content of the Amman temple points strongly to an intrusive population rather than nomads. Much later, it was customary to construct in the open country shrines that were foci of unity for more than one social unit—i.e., neutral territory was deliberately chosen. See A. Laumonier, *Les cultes indigènes en Carie* (Paris, 1958).

place names and proper names in the Bible is admittedly a very slippery business.

According to the last, and now archaic, treatment of the problem, the biblical proper names identifiable as belonging to the Israelites include perhaps one-fourth to one-third which have no etymology or formal analysis stemming from the Semitic language family.[41] If one adds to this the indeterminate number of names which have a "received" popular etymology, but have a more probable explanation from non-Semitic languages of the ancient Near East, there is an impressive percentage of ancient Israelite biblical names which they share, in the first place, with non-Israelites, and, in the second place, with persons of demonstrable non-Semitic origin. It is undeniable that such a thesis will need far more detailed treatment, and also far more evidence than is now possible. Here, it can be only argued that preconceived ideas derived from Byzantine traditions and nineteenth-century scholarship (which had more in common than either knew or could have known) must be subjected to a thorough reexamination.

Because of the sentimental attachment to the more historically important biblical sites of Palestine proper, entire cultural areas of the greatest importance in biblical tradition and extrabiblical history have been virtually ignored. Transjordan and Syria are the classic examples that are crucially important to the present thesis and also to the ancient historical process. We have next to nothing from Transjordan which would give us archaeological controls over our ideas of history in the most important, formative period of ancient Israel. *Any* reconstruction, therefore, is necessarily a venture in prophecy: that future excavations will vindicate the predictions made.

The predictions (*in loco* history) presented here are as follows. With the growing chaos in the north of Syria and the Hittite Empire, the migratory flow from north to south increased during the thirteenth and twelfth centuries. This would correlate with the radical drop in density of population in the northern areas, which seems to be progressive and cumulative during this period.[42] By the end of the thirteenth century and perhaps several decades earlier, centralized despotisms (kingships) arose in Transjordan, after the model so well illustrated from Alalakh, Ugarit, and the various Neo-Hittite city-states. These were not so affluent, nor so "civilized" as those further north, but they illustrate many of the same cultural traits. Since the population pressure is from north to south, the further south, the more archaic the political tradition. Edom is Hurrian, if the usual scholarly

41. Martin Noth, *Die israelitische Personennamen im Rahmen der gemeinsemitischen Namengebung* (1928). Perhaps 20 percent of Hebrew proper names have no adequate explanations, and many others are simply popular etymology.

42. Central Anatolia seems to have been virtually depopulated during the period of 1200–900 B.C.

identification is correct, which we cannot accept uncritically since the same consonants may be correlated with a Canaanite designation of some kind of "nobility."[43] Regardless, Sihon is designated regularly as the "King of the Amorites," which can quite concretely be correlated with the kingdom of Amurru that appears as a *political* designation only in the Amarna and subsequent Hittite sources. With the historical records now extant, it can be traced for some six generations from Abdu-Ashirta, the archenemy of the king of Byblos, Rib-Addi, to Shaushgamuwa, the Luwian end of the dynasty.[44] Presumably it was destroyed by the same vague wave of sea-people that destroyed Ugarit. We suggest that it was in some way transported, in tradition at least, to Transjordan as a last refuge where it survived for a generation without grass-roots support before it was eliminated by the apolitical movement of ancient Israel.

Further north, the kingdom of Bashan was ruled by a semilegendary King Og. Here is a name which can be quite certainly ruled non-Semitic, and identified with both Hittite and Luwian Ḫuḫḫa, and later Lycian *Kuga*.[45] He is repeatedly identified as the "remnant of the Rephaim" in biblical tradition, and this can with little certainty at present be correlated with the *rubāʾum* 'prince'[46] of Cappadocia in the Middle Bronze Age, though it is not possible to build any historical theories upon this linguistic identification in the present state of our evidence. All that can be said is that the identification is in harmony with and furnishes some very interesting suggestions concerning the various uses of the term both in biblical and extrabiblical sources. It is most characteristic of Anatolian kings, like those of ancient China and Rome, that they became gods after their deaths,[47] and the Rephaim are equated with the shades of the dead in later Israelite traditions, as well as in Ugaritic.

The chief god of the Luwians was the storm god Tarḫu, who later, as Jupiter Dolichenus, spread across central Europe as far as Brittany.[48] There is good reason to believe that the deity migrated south as well. We find, for example, at Ugarit the personal name *trǵ(n)ds*: Tarḫu(n)dasi, and a similar

---

43. Cf. Hebrew *ḥōr*, Aramaic *ḥr*; also in Arabic and Ethiopic. It is probably the element common in personal names: *ḥo/urr*.

44. See now the new materials in *Ugaritica*, 5: 113–14, and Schaeffer's discussion on pp. 640–95.

45. For *ḫuḫḫa* and its later forms, see *LPG*, p. 142.

46. But cf. also the Ugaritic name *rpʾan*, and *Dnil mt rpʾi*. The new mythological texts from Ugarit give further compatible evidence.

47. Deification of the deceased king was standard. If Hebrew *rpʾm* can be correlated with an archaic north Syrian root *rpʾ* meaning 'lord,' then the Anatolian custom would account for the biblical identification of the *rᵉphāʾîm* with the shades of the dead.

48. P. Merlat, *Jupiter Dolichenus* (Paris, 1960), pp. 9–24, for the scope, chronology, and process involved.

name recurs in the Aramaic papyri of the Persian period.[49] In Canaanite of the Iron Age also, the name would of course have to be spelled tr^c-, and we have at least two names which probably contain this divine name in Palestine— first the Talmudic tr^cl, Tarḥula, now Tell Deir^{c}alla in the Jordan valley.[50] It is from this place that the famous tablets were discovered in a new script which has been declared to be "Philistine."[51] But the characters have their closest affinities to hieroglyphic Luwian,[52] and the tablet *could* be read as follows, though with such a small amount of material it would be folly to think that the reading is the only possible and, therefore, correct one.

Lšym.mgr.nksny.lymyn
nḥt.'tyn.^{c}ry.ššyn
ḥšn.
["Lašiyum has delivered my funds to Yamiyanu.
I am content. Sixty donkeys have come. Ḥšn."]

The tablet represents, accordingly, a business document, a quit-claim against Lašiyum who acted as agent in delivery of capital against sixty asses. The other documents in the collection of tablets from Deir^{c}alla fit well into the same category. Though most are uninscribed, the characteristic group of seven puncture marks correlates well with seven strokes as witness marks from tablet C (a marriage contract, according to my decipherment, soon to be published) of the Byblos syllabic texts,[53] to the various groups of seven marks which occur so frequently in pre-Islamic Arabic inscriptions. The find-spot of the tablets, in close association with the temple gate, would support the legal interpretation of the tablets.

Of the characters, several have virtually identical forms in hieroglyphic Luwian, and the peculiar deep punch marks have their closest parallel in Cypriote.[54] The affinities with cultures to the north and west support very strongly the thesis that a considerable and probably politically dominant element in the ancient population of *Tar^{c}ula* was Cilician in origin, and formed part of the same southward movement that saw the continuity of the Late Bronze Age Hittite tradition maintained not in Anatolia itself, but in the Syrian region from Hamath to the Anatolian mountains. Similarly, such

49. G. R. Driver, *Aramaic Documents* (Oxford, 1954), no. 5: tr^cds.

50. Representing original Tarḫuna, with a shift from /n/ to /l/ which is well known even in Anatolian languages.

51. Publication of the tablets in *VT* 14 (1964): 377–79 and plate I.

52. The connection with Late Bronze Age Cypriote tablets also cannot be ruled out.

53. M. Dunand, *Byblia Grammata* (Beirut, 1945), pp. 74–75 and plate IX. For the pre-Islamic Arabic, cf. W. G. Oxtoby, *Some Inscriptions of the Safaitic Bedouin* (New Haven, 1968), nos. 18, 36, 65–67, 185 *et passim*.

54. The punch marks would of course be the easiest way to make a circle on a clay tablet.

continuity has already been demonstrated for the western region of the Anatolian coast of the Mediterranean.[55]

If we rule out the idea that these are pure ethnic stocks which are for the first time engaging in a symbiosis with the indigenous (Semitic-speaking) population, as we must in view of the nature of our documentation for the Late Bronze Age at Alalakh and Ugarit, then it follows that personal names of the newly arrived migrants will exhibit the same range of variety, linguistically, as that demonstrably true of their place of origin. If the Semitic form of Luwian, Hittite, Hurrian, and Hattic personal and place names found in the Ugaritic texts are compared with those from Palestine which have no Semitic etymology (or only a popular etymology, which is virtually always the case), we find a very high degree of correlation which cannot be explained upon any other historical basis than migration. In fact, there seems to be a clustering of Anatolian names in the regions of Benjamin, Bethlehem-Hebron, and the upper Jordan Valley and Transjordan. These are precisely the areas where biblical tradition locates the Hivvites and Hittites, the latter at a very early period. The fact that names of persons which originate in north Syria and Anatolia in the Late Bronze Age are found in quantity in the lists of personal and place names in biblical Hebrew constitutes primary evidence for the historical thesis that the Israelite Conquest did not constitute any large-scale displacement of population, but was instead a new social synthesis at the grass-roots level. This was simultaneously a revolution that did away with the various tyrannies, in the smaller city-states at least, and unified the existing population in a new and effective religious community. At the same time, it demonstrates that the conflict was not a racial or ethnic one, for at this early period there exists no linguistic line of demarcation between Israelite and non-Israelite names, other than theophoric. The religious contrast is present, of course, but members of the "tribe" of Judah exhibit such names as Kilion/Kiliyanu (Hurrian), Kenaz (Kunz-Luwian), Eshton/Ištanu (Hattic 'sun'), or, more probably, Uštanu (Hurrian, 'hero'), and so on. "Moabite" names include Ruth/Ruwanda-Ronda-Ruta (Luwian), ᶜOrpah/Ḫurpa-(Hurrian), Mešaᶜ/Mašḫu (Luwian?), ᶜEglon/ Ḫaḫḫaluwan (Cappadocian).[56]

When we turn to place names and other tribal and subtribal designations, we find the same phenomena. North of Deirᶜalla lies Pella = Amarna Piḫil,[57] but Pella, which was the name of the capital of Macedonia in the time of Philip of Macedon, must go back to an original Piḫil or the like, since the loss of the voiced (?) laryngeal has induced a change in the vowel,

55. A. Goetze, "The Linguistic Continuity of Anatolia as Shown by Its Proper Names," *JCS* 8 (1954): 74–81.

56. See further below, "The Midianites."

57. *ANET*, pp. 326, 489. Cf. *IDB*, s.v., for the known history of the town.

exactly as Greek *pyr* presupposes an original *paǵur*, which we have in Hittite-Luwian *paḫura* (and in Semitic *peᶜōr*) 'fire.' The modern Arabic *faḥil* again is a popular etymology, unless there is some semantic connection between it (meaning "charcoal") and the old word for "fire," i.e., *r* became *l*.

On the Transjordanian plateau is located the curious *Lodebar*, "nothing" by popular etymology; but the name is the precise equivalent to later *L/Rondeberras* preserved in Greek, going back to original *R/Luwanda-tapara*, "Ruwanda is Lord."[58] Luwanda becomes *Lod* as *luwazanti* (Ugaritic *lwsnd*, Egyptian *rws*) becomes both Luz and Laiš,[59] while *Ruwanda* becomes Ruth and Rhodes. *D/Tapara* 'lord, governor,' gives us by popular etymology *Dᵉbir*, just as Egyptian transcriptions yield the name *Qiryat-Sōfēr*: city of the *šāpiru* = "governor" (for which, cf. Judges 5).[60] One name is thus a translation of the other. "City of the book" (*sēfer*) is thus again a late popular etymology.

From the Hebron area also comes the tribal name *Tirᶜātīm*:[61] *Taḫ̆huntimu(wa)* = *Tarkondem(os)* of the Greek period. It is a subsection of Jerahme-el, just as *Jerēmōt/Iyari-muwata* is a subsection of Benjamin. Further, such Levite names as *Mūšī*, *Merārī*, *Qīšī*, and others raise the question of whether or not the curious "ethnic" nature of the Levites may not be explained by their pre-Israelite origin as Luwians, who also were evidently noteworthy for their expertise in rituals.[62] The shift from *Luwi* to *Lēwi* is of course exactly paralleled by shifts from *šum* to *šēm*, and *ʾum* to *ʾēm*, and the "ethnic" *Luwi* fell together with the Semitic *lawī*- 'lent, dedicated.' Only *Luwi* can explain the *ē* by "umlaut."

THE MIDIANITES

For a good brief description of the status of knowledge prior to the present work, see G. M. Landes, *IDB*, s.v. All that we know of the Midianites fits very well into the picture of a seasonal transhumance such as that well described

58. *LPG*, p. 131.

59. Albright, *VESO*, p. 50, and "Notes and Comments," *JPOS* 2 (1922): 184–89. Curiously, in view of the Hebrew *rūt*, the Egyptian *Ru-tu-n* could be correct as it stands. For a counter view, see Gardiner, *Onomastica*, 1: 148*–49; his objections seem no longer cogent.

60. Verse 14, *ševet sōfēr* 'marshal's baton.'

61. 1 Chron. 2:55.

62. The original "secular" and warlike nature of the Levites has been a mystery for decades; cf. especially Gen. 34 and 49:5. Note also the role of a Levite and a Simeonite in the episode at Baal Peor. Is Simeon actually Šamḫuna? See now J. Milgrom, "The Shared Custody of the Tabernacle and a Hittite Analogy," *JAOS* 90 (1970): 204–9, for a number of very striking similarities that would strongly support the identification of the "ethnic" Levites as originally Luwian.

by J. Tracy Luke.[63] It is a regular movement, usually within rigidly (if not politically) prescribed limits, of that segment of a village-based society which is dependent upon the raising of livestock as a means of livelihood. Most probably, this transhumance from one pasture area to another is what was originally meant by Hebrew *sāḥar*. If it has anything to do with traveling tinkers, which is quite possible in view of the Beni Hassan tomb paintings representing such a group, it is a transferred meaning from the pastoral original. Again, in view of the unquestioned assumption in recent contributions to the fascicles of the new edition of the *Cambridge Ancient History*, it must be reiterated that this cultural phenomenon has nothing to do with an independent and contrasting cultural adaptation to desert environment such as that of the nineteenth-century Bedouin. In fact, it illustrates well Service's thesis that true nomadism must be *preceded* by some sort of close symbiosis with sedentary society. Pastoralism in ancient (pre-camel) times was presumably an occupational specialty of the younger family members, as indicated in the Cain and Abel story.

It is entirely in keeping, therefore, with the reality of ancient cultures that the Midianites were the main delegation to north Syria to fetch Balaam. Moab is a political designation, while Midian is the designation of a social organization (tribe) which makes up an influential element of the state. There is nothing at all unusual in the fact that a group which is at home in southern Transjordan or the Negeb or both is found pasturing flocks in the Sinai peninsula in Moses' day—it must have been winter time. Similarly, in the Rome of Caesar and Cicero, the affluent landed gentry had shepherds to take their flocks up into the mountains for summertime pasturage—and they are subdivided into those who took wives along with them, and those who did not.[64] It is amusing to read Varro's almost indignant insistence upon the fact that, contrary to the opinion of his predecessors on the subject, there is an important distinction between agriculture and grazing.[65] The matter was a serious issue at that time because grain farming had ceased to be profitable and the small farmers' lands had gone to ruin when they were conscripted to serve in the many wars of Rome. One can see a remarkably similar process in Elizabethan England.[66] Because of economic changes, farming had there also become unprofitable in comparison with sheep raising, probably because of the great rise in demand for textiles. Large estates were "enclosed" for the purpose, and the former tenant farmers joined the ranks of the unemployed, living as beggars, vagabonds, and outlaws. Discharged veterans swelled the ranks of the uprooted, and in addition

63. "Pastoralism and Politics."
64. Varro *de Agri Cultura* II. ii. 9–10; x. 6–9.
65. *Ibid.*, I. ii. 13–14; Introduction, p. 4.
66. For which, see Aydelotte, *Elizabethan Rogues*, pp. 5ff., 71.

the old bands of retainers of the feudal lords were dismissed, often enough, evidently, to continue their previous practices of looting and brigandry, but now without the protection of a powerful overlord. However, it is necessary to emphasize Keen's statement that there does not yet exist any general book of value on the outlaw.[67] Keen's work, however, makes a very strong case for the origin of outlaw bands as isolated groups of resistance to the brutal and catastrophic military conquest of the foreign and contemptuous Normans. The large number of uprooted of Elizabeth's day, on the other hand, arose from the economic and political changes which attended the breakdown of the feudal system. It would seem that the transition from the Late Bronze to the Early Iron Age saw both processes taking place on a much wider scale simultaneously. It is therefore the acme of foolishness to explain it all as a result of nomadic incursions from the Arabian Desert, for which we have no evidence of any population at all during the period in question.

With this background in the social and economic realities of ancient historical process, we may proceed to a cautious discussion of the Midianites and their identity. If we start with the observation that a non-kin, non-residential social solidarity must have a name,[68] and the name is usually derived from some important segment within the society, then a gentilic based upon a place name from which that segment derived is a most likely candidate. The observation of Goetze and others that Anatolian gentilics derive from designations of settlements and geographical regions applies just as well to gentilics in the Bible. There is actually a town, probably of Arzawa in Luwian territory, called *mada*. The Luwian gentilic, which is not attested, would be *mada-wana*.[69] Ugaritic texts offer *mtyn*, which could be related, but in the absence of syllabic writing of the name there can be little certainty. At any rate, the root *mat/d* is very common seemingly in Anatolian languages. In contrast, there is no name in biblical Hebrew beginning with *ma/id-* which has a convincing etymology from Semitic languages, and only the roots *mut-* and *n/ytn* yield names beginning with *mt-*, neither of which can be the basis of *midyān*. On the other hand, the common West Semitic shift from /w/ to /y/ would yield *mad(a)yana* instead of the original *madawana*. In a different region (predominantly the West Bank), the /w/ was preserved as a diphthong: *madaw(a)n(a)* became *mādôn*. Popular etymology would then derive it from a root *DYN*. The original *mad(a)yan(a)* is preserved in Arabic *madyan* and also in Greek. The name ends in Old South Arabic as the designation of a tribe. The form *madôn* occurs in the Septuagint version of Josh. 15:61 where the Masoretic text has *middîn*,[70] and in the list of allies or vassals

67. Maurice Keen, *The Outlaws of Medieval Legend* (London, 1961), p. 225.
68. *PSO*, p. 116.
69. *NH*, pp. 270, 259–60.
70. Cf. the discussion of Landes, *IDB*, s.v. "Midianites."

of Yabin, king of Hazor (Josh. 11:1), where Madôn has a king Yōbāb which can be most probably analyzed as Luwian $(a/i)ya\text{-}pa(m)pa$. The city is listed also among those whose kings were removed (Josh. 12:19). In 2 Sam. 21:20 ʾîš mādôn (Qere; Ketib: mdyn), which has mystified all translators since the Chronicler emended the text to read middâ (whence RSV "a man of great stature"), is correct as it stands, both the Qere and Ketib: a man of midyān/ mādôn. (Curiously enough, both forms occur also as common nouns meaning "strife, contention," but almost exclusively in Proverbs where the original /w/: midwānîm is preserved five times.) The man in question is characterized as having a curious digital structure as well, and together with his three fellows, is described as having been born to "the Raphah in Gath." The raphah can only be variant spelling of rpʾ, whence rᵉphāʾîm, who are similarly located in Transjordan.[71] This solution of a difficult textual problem simultaneously emphasizes the historical complexity of the "Philistine" population, which, like every other large social organization of ancient as well as modern times, was highly composite so far as its ethnolinguistic history is concerned. It also illustrates the universal tendency of ancient populations to designate persons according to their ultimate (known) place of origin even after the passage of a couple of centuries, even though in the case of the man of Madon of Gath, it would be impossible to say from which of the numerous Madon/Midian he derived.[72]

In summary, virtually all of our evidence for the Midianites, like that of the Israelites themselves (notably the Jacob story), points to a fairly recent northern origin of important segments of the population. The mission of Midianite elders to northern Syria to fetch Balaam, as well as the cult of Baal Peor and the cult of the dead which it presupposes, points strongly in this direction, as does also the archaeological evidence already cited. Finally, since the archaeological materials are not conclusive—they could always be explained as the result of trade or cultural diffusion—it is only the linguistic evidence of proper names which can furnish proof of the thesis, and this is no simple task.

It is generally conceded that demonstrations of linguistic affinity of two groups from onomastic material alone is highly unsatisfactory. In the absence of texts, however, there is no alternative. Fortunately, we do have

71. See above, n. 47.
72. A perhaps even better identification of the origin of Midian derives from the place name Midduwa, which is listed among others captured from Mitanni by Suppiluliuma. J. Garstang and O. R. Gurney, The Geography of the Hittite Empire (London, 1959), p. 46, cautiously suggest identifying it with the Neo-Hittite Maldiya/Meliddu, modern Malatya. Curiously enough, Varro, in recommending to his gentleman-farmer friends that they obtain chickens for their farms of the sort called Medice (i.e., Median) observes that the rude people of the hills also call it Melice.

the evidence of the Deir ʿalla tablets which have been dealt with above, and which demonstrate beyond question the legitimacy of seeking northern origins of those names that have no etymological explanation in the Semitic languages. Further, even names that can be explained from Semitic may have Anatolian origin; the criterion for inclusion is simple. If the name occurs, especially if it is common, in Anatolia and northern Syria, the presumption is that the Semitic explanation is a popular etymology if the name is isolated in Semitic. It is a priori probable that most foreign names would be so treated, and would even bring about the necessary sound changes to make the popular etymology plausible. The unfamiliar is always explained by the familiar, and the fact that even very old Semitic names are subjected to a false popular etymology in virtually every case simply underlines the fact that we must expect similarly false etymologies of non-Semitic names. Etymologies are usually immediately convincing or not demonstrable, and there is no simple formula by which to arrive at some nice, objective, scientific proof. If the name in question does not occur elsewhere in Semitic onomastics, it could very well be such a popular etymology. Yet the paucity of our documentation plus the fact that, as in the case of Midian and Cain, even non-Semitic names can easily be perpetuated and diffused widely among a Semitic-speaking population, make proof impossible—as usual. The operating criterion will simply be the a priori assumptions of the scholar in question and his views of the ancient historical process, with some controls derived from known facts of sound changes which are presumed to lie behind the various contrasts in ancient written documents. One has only the context of historical probability upon which to rely, and it would, for example, be absurd to cite a *possible* Anatolian non-Semitic explanation for a name which is common in Semitic, unless there were other historical evidence, of which we have increasing abundance.

As an illustration of the possibilities now available from greatly increased linguistic and onomastic evidence, the following list of names which are either designated as Midianite or closely associated with the Midianites is proposed, at least as a basis for further discussion and discovery.

1. The five kings of Midian (Num. 31:8; Josh. 13:21).
ᵓeⁱwî: No Semitic etymology. Compare Hittite *awiti*- 'lion'(?), and common Anatolian formative element *awa/i*.[73]

Reqem: No Semitic etymology. As a starting point for discussion, note Josephus's statement, Ant. vii "... Rekem, who was of the same name with a city, the chief and capital of all Arabia, which is till now so called

---

73. G. Neumann, *Untersuchungen zum Weiterleben hethitischen und luwischen Sprachgutes in hellenistischer und römischer Zeit* (Weisbaden, 1961), p. 55.

by the whole Arabian nation, 'Arecem,' from the name of the king
that built it, but is by the Greeks called 'Petra.'" The "Greek" name of
course means primarily not "rock" but "rock-cliff," "rock-mountain,"
or the like, and it has no Greek etymology or explanation.[74] On the
other hand, we have the well-attested north Syrian city *Pitru* = Hebrew
P^etōr, the Lycian city *Patara*, and the common noun in Urartean *patara*
'city.' As the semantic equivalent of the very old common Semitic *ṣur*
(whence the city name Tyre), the probability that the word originates
from one of the numerous Anatolian languages is extremely high. At
the same time, we have another of the new very numerous connections
between Transjordan and the north Syrian-Anatolian region.

Further, the name *Reqem* may well be the original name which was
"translated." We need cite only Latin *arx* 'citadel.' The peculiar name
and spelling of the Amarna city Arḫa/Irqata would thus be explained.
The initial *r-* is vocalic and either /a/ or /i/ could be used as a prothetic
vowel. Also, the meaning "rock, cliff" is quite appropriate, just as
ṣura- had already occurred in Old Amorite as an element in the onomastic
tradition as a designation of god, both in name sentences and as hypo-
coristica.[75] The final *-m* could derive either from Luwian or Old West
Semitic. My colleague, Gernot Windfuhr, kindly informs me that a
word occurs in a modern Iranian dialect which in form and meaning
can be linguistically identified with this ancient name. Similarly, English
"rock" has no ancient explanation. The name *Reqem* occurs elsewhere
in the lineage of Hebron, which the traditions insist correctly was
Hittite in pre-Israelite times, in the lineage of Machir in Transjordan,
whose mother was "Aramean," i.e., (north) Syrian, and in the lineage
of Benjamin, the territory and onomastics of which is full of Syro-
Hittite and Luwian associations. Compare also the Judean place name
Rekah, 1 Chronicles 4:12.

We may compare Rhakius father of Mopsos in early Greek
tradition. *Reqem* as the name of a Midianite king then can be tentatively
identified as the root with the very common suffix element *-m* characteristic of Luwian and other Anatolian names.[76] According to the
same Greek tradition, Mopsos/Muksas went south as far as Ascalon,
and in biblical onomastics, *wopsi* with the well-known *m/w* shift, is
the father of Naḥbi, one of the twelve "spies" sent out by Moses,
mibsam "son" of Ishmael, with the shift from *u/i* and the *-m* suffix
which could be the Lycian/Luwian gentilic ending. Muksas may be

74. Frisk, *Griech. Etymol. Wörterbuch*, p. 523.
75. *APN*, p. 258.
76. *LPG*, pp. 181–82.

identified in the otherwise unknown and inexplicable place name in the vicinity of Beth-Shemesh, Maqaṣ, with a shift from *qs* to *qṣ*, which is regular in West Semitic.

Ṣūr: Common Semitic, from Old Amorite on.

Ḥur: Common formative element in most eastern Mediterranean languages, Semitic and non-Semitic.

Rebaᶜ: Compare Hurrian *arpiḫe* very common at Nuzi.[77] Ugaritic *arpḫn*(?).[78] The initial vowel of Anatolian names beginning with *ar-* is regularly dropped. Since it is most probable that the *ar*-spelling is merely the graphic representation of a vocalic /r/, there is no significant sound change involved. The cuneiform /ḫ/ is fairly frequently rendered by Semitic ᶜ*ayin* = Ugaritic /ġ/, but the material is as yet too sparse to permit systematization.

2. Place names and personal names designated Midianite or closely associated with them.

Kozbī: Luwian *Kunzum-piya* = Greek *Kozapeas* (Num. 25:15).

Peᶜōr: Luwian *paḫ(u)ra* (Num. 25).

Bilᶜam: Compare Ugaritic *plġn*. Could it be Cappadocian *palḫ-*? Note the shift from /n/ to /m/, as in *Gēršōn/m*.[79]

Belaᶜ: Same, without *n/m* suffix.

Bālāq: Same, with ᶜ/*q* interchange which is common in Aramaic. Compare the *ḫ/q* interchange also in Luwian.[80] Compare also *Pa-al-la-qa-a-a*, *Pillako(as)*, and *Palakoas* (uncertain).[81] Compare Houwink Ten Cate, where -*kuwa* is cited in a number of names, and it is suggested that it may be derived from *Que/Hume*.[82] Again, there are too many possibilities for explanation from Anatolian languages and dialects, and only forced if any explanation from Semitic.

Mōᵓāb: Luwian **muwa-apa*. Compare *apa-muwa* and *Obamoutas* = *Uba-mu(wa)ta*.[83] But compare also *Moabis, Moas*.[84] Like the -*kuwa* names mentioned above, the two cited here come specifically from Termessos

---

77. *NPN*, p. 31.

78. *PTU*, p. 365.

79. *Ibid.*, p. 312. *NH* no. 914, *palḫa/uzia*.

80. *DLL*, p. 135. Note GN *jebel balaq* south of Mārib in Yemen. Professor Ghul informs me that the word means "alabaster."

81. *NPN*, p. 110; and *ZPK*, nos. 1257 and 1194.

82. *LPG*, pp. 152–53.

83. *NH*, no. 99 (Tarsus); *LPG*, p. 169.

84. *ZKP*, nos. 940–43, 940–41.

in Pisidia. The biblical explanation of the name is a classic illustration of a story inspired by a popular etymology of a strange name.

Rūt: (Ruth) Ugaritic *rt.*[85] *^{d}Ru(wa)(ti)a-s,* the name of the Cilician stag-god in hieroglyphic Luwian still worshipped in the Hellenistic period in Cilicia and Cilicia Aspera, and identified with Hermes, as Tarḫu was identified with Zeus. This is precisely the situation presupposed in the Acts 14:8–18 narrative in which Paul and Barnabas at Lystra were identified (in the Lycaonian tongue!) with Zeus and Hermes. Note also the nearby place *Derbe* and compare *Rōn-derbe-mis = Ru(n)-tarpa + mi.*[86] Lystra could well be *Luz/s-* plus the unexplained *-tra* which is fairly frequent in Anatolian names such as Moatra, Pinatra.

Ṣippōr: "Father" of Balaq, king of Moab. (Cf. *Ṣippōrâ,* daughter of Jethro, priest of Midian, and wife of Moses.) No other names in biblical Hebrew with this root. Compare *ṣpr, ṣprn,*[87] but also the Karatepe bilingual which equates Phoenician *rśp ṣprm* with hieroglyphic Luwian *^{d}CERF-ī-ś-(ḫa).*[88]

Yitrō: Old Amorite *yatar- +ā?* Compare *yariḫā = y^{e}rēḫō* Jericho. Shift from *yatar* to *yitr* is exactly parallel to the shift from the older *malik* and *ʿapir* to *milk* and *ʿibr,* respectively.

Ḥōbāb: Compare Hurrian *Ḥumpabe, Ḥupabe.*[89] Ugaritic *ḫbb.*[90] It is not likely that the name is derived from the Semitic root ḤBB 'to love,' since this word does not occur in early sources. Deuteronomy 33:3 derives from the ḤBB underlying the Mari *ebēbu* 'purify,' and the same root could explain this name. However, the great frequency of the Hurrian Anatolian names with *ḫu(m)p-* combined with the other evidence for Anatolian names would give this the preference.

ʿēpâ: (Gen. 25:4). Compare Ugaritic *pdġb* which can only be Hurrian *pudu-ḫeba(t).* Cf. *abdu-ḫepa,* king of Jerusalem.

ʿēper: (*Ibid.*). Probably Old Amorite (or Canaanite?) name occurring in the Execration texts as *ʿpr-* and APN *ḫapir-.* According to the present thesis, it would have nothing to do with *ʿpr = ʿibrî,* but with the root preserved in Akkadian *ebūru* 'harvest.'

Yoqšān and *M^{e}dān:* (Gen. 25:2). It may be nothing more than a curious coincidence, but the names correspond to the name of the capital city and the country of Wašukanni and Mitanni, respectively, with meta-

85. *PTU,* p. 312. Cf. the discussions, *LPG,* pp. 128–31, and *HH,* p. 102.
86. *LPG,* p. 161.
87. *PTU,* p. 413.
88. *HH,* p. 63.
89. *NPN,* pp. 63, 217.
90. *PTU,* p. 387.

thesis in the first name. In view of the pronounced Hurrian elements in early biblical onomastics, there is nothing particularly surprising about the survival of old political names as tribal designations in a different geographical area.

This list merely illustrates from a few examples the high degree of probability for the Anatolian interpretation of many names from throughout Palestine and Transjordan. The major problem of course is to discover what happens to exotic and strange names when they are introduced into a predominantly Semitic-speaking area—and transmitted for decades, if not centuries, before being reduced to a more or less fixed written form. The picture is just what we would predict from what is known of social structures and historical processes of the Late Bronze and Early Iron Ages. A mixed, foreign, non-Semitic population moved south, superimposed itself upon and intermarried with an existing population, the names of which can largely be traced back to the dawn of the history of West Semitic, i.e., to the earliest recorded names in Egyptian and Akkadian sources. The result was a new synthesis; a few generations later, the problem of origins was insoluble. The genealogies preserved only memories of local political relationships which were constantly changing, and of some association with the north.

Finally, the question may be raised of whether the *name* *Madyan may not be closely related in some way to that of the Madai, Medes. As long ago as Herodotus, the association of the Medes with the Ashkenaz/Scythians was a fixed tradition. On the other hand, the Table of Nations in Genesis 10 very closely associates Gomer (Cimmerians), Madai (Medes), Javan (Ionia), Tubal (cf. Tubal-Cain = Tu/abal), Meshech (Mushki-Phrygians), and Tiras (Tursa/Tyrrhenians = Etruscans). Midian is lacking, while Ashkenaz (= Scythians) is derived from Gomer. Midian does not seem to be referred to in any historical context after the event of Judges, chapters 6–8 (cf. 9:17), but the Kenites and Amalekites remained historically very much alive until the former were absorbed into Judah and the latter evidently disintegrated as a social organization under the blows of Saul and David. The only way in which historical sense can be made of the evidence we have is simple: merely to recognize the fact that the "gentilics" designate social organizations and their successive continuities, not races, and both theoretical as well as historical data furnish very sufficient grounds for the observation that such large social organizations, including politically organized states, were very short-lived in ancient times, as well as more recent ones.

It goes without saying that far more refinement needs to be brought into such comparative studies, which range from the Bosporus to Nineveh and from the Black Sea and beyond to the Arabian Desert. But it must be remembered that this is only slightly greater than the actual range of occurrences of hieroglyphic Luwian inscriptions between 1500 and 700 B.C., when they seem finally to have become extinct as a written language system—probably in favor of the Semito-Greek system which soon developed throughout the Mediterranean. The traditions of Genesis, chapter 10 exhibit considerable familiarity with Anatolian population groups, and King David had covenant and doubtless commercial dealings with Toʿi, king of Hamath.[91] King Solomon received his charisma of wisdom at the high place of Gibeon whose god must therefore have been identified with Yahweh.[92] It is that wisdom which, like its modern counterparts, soon found itself involved in a bitter struggle against the older religious tradition. Wisdom as a technical (relatively speaking) tradition was useful for gaining immediate goals, and if it enabled Solomon to solve the case of the harlot's children, it also enabled Amnon to find a way to rape his half-sister.[93] Technology has no built-in self-control, and the human engineering that accompanied the rise of empire and centralization of power was unable to contemplate that ignoring ethical controls might have serious adverse side effects. Whatever else it may be, ethic is a predictive factor, because of the fact that there is a limit to human beings' toleration of those conditions imposed upon them by superior force that reduce them to mere things and make life unbearable.[94]

The contacts with ancient Anatolia are constant, therefore, which explains much that was formerly inexplicable in biblical history. However, it must be pointed out that Anatolia itself by no means sprang full-grown from the head of Zeus, Santa, or Tarḫu. As Goetze has pointed out, the Hittites seem to have had virtually no higher culture at the time they came into contact with the older civilizations of the Near East.[95] No society has pulled itself by the bootstraps from barbarism to civilization without the aid of the accumulated experience of humanity:[96] indeed, this may be the

---

91. The name is, of course, the well-known *Tagi*, probably Hurrian. *NPN*, pp. 261–62 (2 Sam. 8:9–10). Could it be *Taḫe* instead? Cf. *ibid.*, s.v. *taḫ*. If the root is *tak*, the name would be the semantic equivalent of the Semitic *yākîn*.

92. 1 Kings 3:4–15. Cf. 2 Sam. 21:2. There can thus be no doubt that at least some of the "high places" so bitterly condemned, especially by the Deuteronomic historian(s), were actually survivals of pre-Yahwistic cults whose gods were identified with Yahweh under the Monarchy.

93. 2 Sam. 13:3. Translations obscure the point by rendering *ḥākām* as "crafty" (so RSV), instead of "wise."

94. Cf. the universal rise of ʿApiru movements, chap. V.

95. *Kleinasien*, pp. 171–78.

96. On the diffusion of civilization from Mesopotamia to China, see Carl W. Bishop: "Mainly therefore, it would appear, to the stimulus imparted by cultural diffusions from the ancient Near East must have been due the origin and fundamental type of that civilization which eventually took form in Eastern Asia" ("The Beginnings of Civilization in Eastern Asia," *JAOS*, suppl. no. 4 [1939]: 61).

important aspect of civilization itself. To reject civilization, particularly in the area of humanity, to throw out as worthless all the preserved experience of human beings concerning those things that comprise peace, is merely to turn the clock back to the barbarous splendor of the Late Bronze Age which ended in catastrophe. It is too expensive in the coinage of human life and human misery. Although the biblical revolution showed another way, it seems that humanity has seldom wanted to predict.[97] But it is not *necessary* to experience the destruction in order to learn the things that make for peace.

Misery knows and knew no ethnic boundaries, and the two–century long harrassment of those who were not politically important illustrated by the Amarna letters on the one hand and the early biblical tradition on the other prepared the way for a religiously based movement of unity against a complex of power structures that had long outlived their usefulness. The powers had been more interested in competing with one another for control than in functioning for the well-being of human beings. If the kings of the Late Bronze Age regarded their dominions as something delegated to them from the divine world,[98] it needed only the introduction of an ethic to see that the divine world itself could rule without the extravagantly expensive prestige symbols of the temple, palace, and military establishment of the kings. The Mosaic covenant provided this ethical system, and it created a new people out of the ashes of the Late Bronze Age cultures. The history of Western man is a history of the alternation between the ethical principle and the technological-political one. This is the battle between Yahweh and Baal:[99] the Lord of the All Powerful State, the source of all prosperity and security, the Lord of Heaven and Earth, but actually the ancient monument to the primitive tribal mentality which was at least for a short time abolished and transcended in the Mosaic monotheism of ancient Israel. But though Baal dies, he perpetually rises again.

97. The principle of prediction as the basis of ethic is well known in the so-called "golden rule."
98. Frankfort, *Kingship and the Gods.*
99. I Kings 19:18.

VII

TRIBE AND STATE IN THE
ANCIENT WORLD: THE NATURE
OF THE BIBLICAL COMMUNITY

It has been assumed almost unquestioningly by traditional academic biblical scholarship that the tribal structure of early Israel constitutes proof that it was both primitive and nomadic.[1] The tribal organization is presumed to have been a temporary hold-over from a desert nomadism, which was soon outgrown when, in the course of evolutionary development from primitive culture, the tribe eventually became civilized enough to become a state. Though this evolutionary theory of biblical history increasingly has come under attack as well as defense, it seems to have escaped the notice of biblical scholars that the whole concept of social evolution has undergone most radical changes since the nineteenth century (even though some anthropologists still have nostalgic attachment to nineteenth-century theories).

After nearly a century of increasingly sophisticated ethnological description since its beginnings as a "hobby of playboys and missionaries," the evolution of cultures and societies is no longer an eccentric branch of philosophy, but an ordering of social and cultural systems, a classification of types in an ascending scale of complexity and access to energy sources, of specializations and multiple relationships.[2] It is seen especially that beyond the truly primitive band of food-gatherers, kinship is rarely functional in the formation or preservation of large social units.[3] The history of large units which actually do have kinship ties is one of both fission and fusion—and I would suggest that considerable social and emotional energy is exhibited when either takes place in a given social field.

As E. R. Service has said, "Social structure, as a matter of fact, would seem to be the result of the workings of other factors and the cause of

1. So, most recently de Vaux, *Ancient Israel*, pp. 4–12.
2. *PSO*, pp. 5, 127, 131.
3. *Ibid.*, p. 112.

174

nothing."[4] Those "other factors" constitute the historical problem in accounting for the existence of that large social organization known as the twelve tribes of ancient Israel. As a matter of fact, they appear to have been the largest social organization known in the latter half of ancient Near Eastern history which was not based upon a monopoly of force in a political state. Although tribal organizations tend to be ephemeral in comparison with political states and empires,[5] nevertheless in important respects this period of ancient Israel's history remained normative for many centuries. It laid the foundations for a cultural continuity with a tenacity which outlived the supposedly much more efficient and indubitably more wealthy and powerful politically organized cultures of the ancient world.

It would be valuable therefore to examine the ancient social organization in the light of this fact and also in the light of the new methods and data stemming from the renewed interest in social evolution based upon direct observation rather than on a priori evolutionary theories. In other words, evolution is a conclusion based upon observation, and not a ready-made category into which to force what little data we have or a theory which enables us to assign a biblical passage to one or another of the supposed phases of evolutionary development which the ancient community is thought to have passed through before it became "civilized."

One of the most striking facts which emerge from a study such as that of Service is the observation that none of his categories from the lineal band to the chiefdom or primitive state furnishes a classification into which we can place either a particular Israelite tribe or the tribal federation as a whole. Though there are certain facts which "fit" and are illuminated by the modern anthropological description of primitive society[6] (which should not be disregarded), yet never does one have the impression that any discussion of a particular social structure is what enables us to classify and understand the functioning of the ancient tribes of Israel. It is not surprising, since the field

4. *Ibid.*, p. 180.

5. *Ibid.*, p. 114. Most states of the ancient world were equally short-lived.

6. Many of the observations fit much of human history, ancient or modern: e.g., "The external policy of tribes is usually military only. Usually, too, the military posture is consistently held; that is, a state of war or near-war between neighboring tribes is nearly perpetual. Tribal warfare by its nature is inconclusive. . . ." Compare the thousand year hostility between East and West from Darius to Muhammad, or the "cold war" in more recent times. The point is that political policies (i.e., policies of supposedly civilized states) can too easily be subsumed under this same observation. Service here is rather naive in failing to observe the function of ideology in both creating and sustaining the "military posture," or at least in concluding that "terrorization, or psychological warfare, seems to be at its highest development in tribal society" (*ibid.*, p. 115). A daily newspaper will furnish ample illustration; compare the foreign policy of the Hittite Empire discussed above.

is still emergent and restricted to the modern world in which nonpolitical societies are both limited in their scope of action and strongly influenced by civilization. This observation, however, underlines the situation in the ancient world. It is now inconceivable that there could be any large population segment in the thirteenth century B.C. anywhere in the fertile crescent that was not under a powerful influence—political, cultural, or economic—from the network of empires which divided the fertile crescent among themselves. To this we may add the insight that as social units become larger, kinship ties become increasingly dysfunctional as the basis for the larger group; but kinship terminology seems to become more used to express the new bond that ties the larger group together.[7] With the state (as with the primitive chiefdom), closeness to the putative common ancestor becomes of great importance for establishing rank in society.[8]

Though none of the categories of band, clan, tribe, chiefdom, or primitive state seem to fit any of the larger segments of ancient Israel, still some of the operating characteristics in the social evolutionary pattern described by anthropologists can be seen to be true of ancient societies. Since survivals of very ancient traits seem to be constant in long-standing cultures,[9] we must ask whether or not those primitive traits which can be seen in ancient Israel are not, as a matter of fact, either survivals or deliberate revivals of already moribund cultural institutions which received new life because of the usual multiplicity of factors.

Once the true lineage band of primitive society gives way to larger,

7. "It is noteworthy that many of the sociocentric status terms in tribal society are derived from kinship criteria" (*ibid.*, p. 131).

8. "Persons are then ranked above others according to their genealogical nearness to him [i.e., the chief]" (*ibid*, p. 155). As the narrative of Solomon's accession proves, genealogical nearness to the putative "common ancestor" had absolutely nothing to do with the resolution of the purely political conflict over the succession to the throne. It was a matter of pure (!) politics, won by the side that had nothing to do with the old Mosaic religious tradition. Bath-sheba, Zadok, and Nathan all stemmed from the old Jerusalemite political-religious aristocracy, just as the political organization under Solomon was a continuation of the pre-Davidic, Canaanite (or should we say Syro-Hittite in view of Ezek. 16:3?) city of Jerusalem whose pre-Jewish history is attested for a thousand years before King David. As in the case of Samson's enemies, the parable of Nathan shows also that even the ancient pagans could recognize a gross misuse of power. If anything *could* have been said about the Israelite background of Nathan, Zadok, and Bath-sheba, such information would have been included in the narratives and genealogies. It has for decades been recognized that Zadok's genealogy in various places in the Bible is a record of succession to a priestly social position that has nothing to do with kinship and, in the later stages, little to do with the Mosaic tradition.

9. Compare the institution of the "Queen's Bed-jumpers" in contemporary England, or, on a more relevant level, the increasing popularity of appeals to tribal violence as a remedy against politically legitimized decisions and procedures.

more efficient, and more complex forms of social organization, relationships of individuals likewise become more complex and multilateral. The position of the individual is virtually always that of the center of a series of overlapping and concentric circles. If he is a member of a family, he is also a member of a larger group which may or may not actually be held together by traced lineage bonds. On the higher level, he is a member of a tribe in which it is virtually inconceivable that traced genealogy is functional as the basis of tribal solidarity. On the next higher level, the federation of tribes represents the largest social unit in which the individual has status, but in the Hebrew Bible no *individual* is ever designated or identified by the name of the largest group (with one exception from a source which all scholars have agreed is among the latest in the Pentateuch),[10] i.e., "Israelite."

Each of these circles of social organization has its own structure and function—and they are often in competition, for obligations to the lineage may well conflict with obligations to the tribe or the federation, as many biblical narratives of the early period well demonstrate.[11] Each of the circles has its own kind of leadership and its own characteristic functions—and these of course change as the social structure changes, as it did radically at least four or five times in the course of biblical history. It is for this reason alone that a purely formal analysis of ancient culture can yield neither good historical description nor any adequate understanding of the dynamics of historical change and continuity. Form and function are not inseparably bound together: a form of organization or behavior may serve opposite functions in different social contexts. *Si duo idem faciunt, non est idem.* [If two do the same thing, it is not the same.]

KINSHIP AND SOCIAL ORGANIZATION

No one is likely to deny the constant importance of real kinship in ancient Near Eastern cultures, or for that matter in most societies of human history. Nevertheless, the function of real kinship ties in society is so limited that something larger is needed, particularly as population density increases and social conflicts become more complex. Furthermore, kinship ceases to be of much importance as the common ancestor of two persons becomes more remote. The five-generation pattern, according to which two individuals are related if they have a common ancestor within the fifth generation,

---

10. Lev. 24:10–11.

11. The most dramatic narrative of this sort is preserved in the story of the rape of the Levite's concubine (Judg. 19–20). Another is preserved in Josh. 22:10–34, although it has clearly been reused centuries later than the events recorded for other purposes.

seems to have been characteristic of the early biblical community.[12] It would be absurd, however, in view of the great diversity which characterized the twelve tribes, to maintain that this pattern must have been true of all segments of the federation. But the Decalogue itself (or a subsequent addition which still must be early) indicates in its "statute of limitations" that the consequences for radical evil committed by an individual may fall upon his descendants until the third or fourth generation. Beyond the fourth generation of descendants there is no corporate responsibility, and probably no other important social function. Forms of social organization not based on kinship have already taken over in the earliest recorded historical societies.[13]

The history of society is one of both fission and fusion. That the biblical culture was quite aware of fission as the source of new social units is demonstrated by all kinds of narratives. The Tower of Babel story is the parade example, followed by the Table of Nations (Genesis 10), in which the independent existences of all sorts of cultural units are described as a genealogical descent system—but based upon cultural affinity and political history rather than real ethnic, much less genealogical or even linguistic relationships. The separation of Abraham and Lot is another good example, as are the separations of Jacob and Laban, Jacob and Esau. The process is beautifully illustrated in the "genealogies" of Chronicles, which combine the complexities of local tribal history with observations concerning the lineages of particular persons in such fashion that it is doubtful that we can ever determine where the former ends and the latter begins. For example, Ḥur in 1 Chronicles 2:19 is the son of Caleb by his wife Ephrath. Caleb in turn is the son of Ḥeṣron, and grandson of Pereṣ. In 1 Chronicles 4:1, Ḥur is a son of Judah together with Pereṣ and Ḥeṣron, and Caleb has disappeared entirely from the genealogy, but Ḥur is still the "son" of Ephratha the "father" of Bethlehem.[14] There is much authentic social history preserved in the genealogies of Chronicles, if the underlying systems of thought upon which they are based can be recovered. The early ones have little if anything to do with real kinship systems, but are the record of all sorts of tribes and clans which constituted the extremely complex social structure of early Israel and of Pre-Israelite

12. Exod. 20:5.
13. The historical and linguistic evidence is sufficient to prove this statement so far as early Israel is concerned. See chap. V.
14. Probably what is involved as well as tribal politics in such genealogies is the process that anthropologists call "swarming"—a segment of a village moves out and establishes another village. In time the immediate genealogical connection is denied, and the new village establishes its direct connection to a more remote tribal "ancestor." Both the sexual mores and even gender of "ancestors" are made to submit to the social concerns for establishing the genealogical proximity to the common ancestor as an expression of the constantly changing prestige system of parochial, tribal politics.

populations of the region. They have nothing to do with "ethnic identity" since such a concept was virtually nonexistent in ancient times, though unity was constantly a political goal. The means was simple; virtually all political relationships not based upon the frank imposition of a domination of superior power were described in kinship terms—to such an extent that we are often hard pressed to determine whether one king who is described as the "son" of another king is actually his physical descendant or merely his dependent vassal. The constant problem faced by ancient (and modern) large social or political organizations was to transfer the allegiances of the small kin group to the larger social organization. The same problem was already present in very primitive "tribes." In a clan organization as described by Service, it was regarded as a violation of decency to refer to kinship at a meeting of the clan, since real kinship ties were likely to disrupt or unnecessarily to polarize the members of the larger group which had at least for a time transcended the blood tie.[15]

If the tie of blood was not the real basis for the larger social group, what was? The concept of "race" may be dismissed from discussion as a relatively modern invention, and one which the human race can well dispense with in the context of social organization. But it needs to be considered whether or not the concept of race is the modern correlate to the ancient concept of the legendary or mythical common ancestor. In both cases, the major social purpose of the concept is the concern for the continuity of some existing or emergent social contrast for political purposes—above all, for preservation of the group—and the guarantee that the conflicts internal are less important than those external.[16]

Much of recent anthropological study has tended toward the conclusion that the major function of non-kinship-based social organizations is cultic/religious in nature. I would agree most emphatically with Service in concluding that the function of clans is not religious liturgies, but the opposite: liturgies are the major concern of the clans since it is this which holds them together, and little else[17] (unless there is real estate involved). It is no doubt controversial that this is an adequate explanation of religious communities. I would insist most emphatically that no understanding of the Old Testament or the New Testament communities can be obtained historically from such a point of view, though it is most useful for reaching a conclusion concerning the postbiblical Hebrew, and the post-New Testament communities involved.[18] What we are concerned with is a consistent transition from an

15. *PSO*, p. 117 and n. 2.
16. So, *ibid.*, p. 114.
17. *Ibid.*, p. 126.
18. That there were very similar developments even in pre-Exilic Judah can easily be inferred from the prophetic denunciation of such "ceremonial labors."

ethical/religious value system to a political or social class system in which the inherited religious tradition is used as a justification for the rejection of legitimate claims by persons who then have nothing to gain by continued identification with the political/religious power structure.

For much the same reason, societies are very reluctant to admit the process of fusion as the historical reality underlying their sociopolitical existence. Any emphasis upon the diversity of social origins is likely to reinforce the present stresses upon the existing newly formed community which is likely to be harassed by internal stresses deriving precisely from that diversity. Therefore, the diversity will be tacitly ignored in all official records and traditions, in favor of the common ancestor or common tradition which is the common bond between them.[19] It is no wonder, therefore, that when the "common ancestor" tradition was appealed to in biblical tradition after the conquests of King David, it was necessary to go all the way back to a period five centuries or more before the time of Moses. But it is not demonstrable that this "common ancestor" tradition was of any importance whatsoever in biblical tradition prior to the time of Solomon, and then it was introduced for specifically political, not religious, reasons. Early traditions such as chapter 24 of Joshua indicate the contrary: that a commitment to Yahweh as the basis for the new community excluded the possibility of commitment to the pre-Yahwistic tribal or ancestral cults which derived from "beyond the river." That there was a concept of a "common ancestor" prior to the rehabilitation and assimilation of the Abrahamic tradition during the United Monarchy is demonstrated by the traditions concerning Jacob and the identification of Jacob with Israel, which must have taken place very early after the formation of the federation. But the fact remains: the Abrahamic tradition became socially functional only after the incorporation of the old Canaanite cities into the empire of David, and, as the prophetic movement soon found, the result was a thorough paganization of the state and culture. The biblical revolution would have ended with Solomon, were it not for the fact that the paganization resulted in the same intolerable social conditions that underlay the biblical revolution in the formative period of Israel. Revolution was followed, as usual, by counter-revolution. We have an almost perfect historical parallel for the process in the history of Zoroastrianism. The establishment of the religion, at least under Darius, meant an inevitable political reception of certain features of the old pre-Zoroastrian Iranian paganism,[20] since the state had to recognize the religious diversity of the Iranian populations, only part of whom were actually Zoroastrian religiously.

19. The same tendency may easily be observed in modern church mergers.
20. I. Gershevitch, *Avestan Hymn to Mithra* (Cambridge, 1959), pp. 13–22, describes the process.

The myth or legend of a tribal ancestor is common to a tribal form of social organization among very many modern primitive groups, and is certainly extremely ancient. It has been frequently observed that a most interesting pattern exists in the genealogies of the Bible. In the description of lineages a series of personal names characteristically ends with the name of a tribe—and, with perhaps one or two exceptions, none of the genealogies carry back to the period before the Conquest. If early Israel were actually the continuity of nomadic tribes going back to Abraham, this is contrary to what one would expect. It is most tempting to see in this genealogical discontinuity which coincides with the date of Conquest a reflection of the social fact that persons received a new social context, a new identity, which virtually wiped out the older social structures which must have existed in great variety. In these circumstances, for the first time, the social function and necessity of the stipulation to honor thy father and mother—rather than remote ancestors—become comprehensible. I would suggest as a corollary that, like the rest of the commandments, it is easier to promise than to obey, and the older tradition survived in the cult of the "high places," which became increasingly incompatible with ancient Yahwism, *especially* after the political establishment of Yahwism under the Monarchy. For the assertion of local, particularistic cults against the established "religion" of the Jerusalemite political authority meant also the rebellion against those aspects of the inherited faith that were responsible for the creation of the community in the first place, particularly the religious ethic, to judge from the prophetic condemnation. The exploitation of the religious tradition for the purpose of maintaining political solidarity and power was thus self-defeating, for the process was thoroughly pagan. The idolization of power could only hasten the fragmentation of society, setting even family members against each other. The result was juvenile delinquency and women's "liberation": "As for my people, children are their oppressors, and women rule over them." It is significant that Islam very early in its history likewise considered the abolition of local tribal mosques for precisely the same reason.[21]

The social diversity of the population of Palestine during the federation is illustrated by the very diversity of names for the smaller units. There is little doubt that important lineages were so included, but the extraordinary frequency of place names as gentilic designations for smaller social units demonstrates the complexity of even a single tribe. The dialectal distinction between Gilead and Manasseh/Ephraim illustrated in the *šibboleth* incident points in the same direction, as does also the peculiar land tenure-inheritance custom of the towns identified as the "daughters of Zelophehad." The custom, which must have been in existence for some time before the

21. Qur'an Sura 9: 108–9. *The Encyclopedia of Islam*, 3: 318–20.

"Conquest,"[22] contrasted to the normal mode of reallocation of family holdings under Joshua and Moses according to the tradition, which we have no reason to question. Under the circumstances, the act preserved in several traditions can only have been the "recognition" of existing custom, since it did not conflict evidently with the religious-covenant basis of the community.

Finally, the very fact that the names of all of the tribes are subjected to popular etymologies which are patently false in every case demonstrates the great antiquity of the names and the *dis*-continuity of tradition which was a major social function of the new religious community. The continuity of the formal name, however, demonstrates also the long-standing Palestinian origin of the populations whose social organization changed so radically and so rapidly at the time of the Conquest.[23] A common feature of new social organizations is the assumption of the name of an originally important lineage or tribal segment within the whole. As in modern church mergers, the new organization is regarded as the legitimate successor to all segments and forerunners. It is this which is reflected frequently in legendary genealogies, and which is most beautifully illustrated in the recently published genealogy of the dynasty of Hammurapi,[24] as well as the Assyrian king list.[25] It seems extremely probable that there was at one time a tribal political organization designated by the name of Abraham and no doubt Isaac as well. The tribal organization evolved (exactly as at Byblos) into a political state in the Hyksos period, and the Abraham legend became politically functional at Hebron precisely as the Keret Epic (and others) were in Ugarit.[26] Not until the reestablishment of kingship at Hebron was the old political legend again valuable as a means of ideological unification of a diverse population.[27] It is one of the ironies of history that it eventually became a symbol to support the opposite. At any rate, there can be little doubt that the major function of the narrative was to indicate that David as king represented the legitimate

22. In view of the pronounced Luwian concentration in the Shechem region, it is perhaps more than coincidence that Herodotus describes as unique and peculiar the matrilineal customs of the Lycians, who are certainly a Luwian population.

23. For the antiquity of the names, cf. W. F. Albright, "Northwest Semitic Names," *JAOS* 74 (1954): 222–33.

24. J. J. Finkelstein, "The Genealogy of the Hammurapi Dynasty," *JCS* 20 (1966): 95–118; A. Malamat, "King Lists of the Old Babylonian Period and Biblical Genealogies," *JAOS* 88 (1968): 163–73.

25. A. Poebel, "The Assyrian King List from Khorsabad," *JNES* 1 (1942): 247–306, 460–492; 2 (1943): 56–90. Cf. the discussion of B. Landsberger, "Assyrische Königsliste und 'dunkles Zeitalter'," *JCS* 8 (1954): 31–45, 47 73, 106–33, especially pp. 31ff. and 109ff.

26. Just what that political function may have been we are in no position to say.

27. It seems very probable that the United Monarchy under David and Solomon carried out a consistent policy of unification of the population by playing down the old inherited religious contrast between Israelite and non-Israelite.

successor to the old pre-Israelite dynastic tradition, probably of the Hyksos period.[28] Again, the whole process would have been meaningless had there been a complete discontinuity of population groups in the south as the result of the "Conquest."

Though study along these lines is obviously in its infancy, such an approach demonstrates the usefulness of the new evolutionary studies, now several decades old, in anthropology. Among other things, we can see almost perfect correlations with very early historical traditions in the Near East. At the same time, it is quite clear that by the time we have written documents to check speculation, the category of primitive society has already become irrelevant. What we are dealing with are certain formal elements which are often remarkably persistent long after society has emerged from a primitive type of economy, social organization, and ideology. The forms survive not because of any intrinsic meaning—they are always, like old names, re-interpreted in the light of the present—but because of their social utility in maintaining the solidarity or the functioning of the large social organization.

As an example we may cite the Execration texts. It has often been observed that they demonstrate a transition from a tribal sort of social organization to a chiefdom[29]—two of the important categories according to the evolutionary theory of Service. However, it is most probable that in the Bronze Age such evolutionary developments were not one-way streets. The process was reversible, as it was also in the Mycenean-Minoan culture evidently. And ancient Israel is the best documented—and least understood— example of the process. The religious ideology of ancient Israel stands in constant relationship to the form of social organization—as one changes, so does the other, and it is this, plus the fact that there never was an ideological unity until the post-Exilic period,[30] that makes the history of ancient biblical faith so incredibly complex. Yet, to attempt to describe that faith apart from its functional relationships to and in the concentric circles of social organiza-tion is merely an exercise in anachronism—of use to the contemporary community of the scholar who engages in it, perhaps, but of no utility in understanding what that faith actually was, how it operated in culture, and why it had such remarkable tenacity.

If religious faith is thus intimately bound up with social organization in the Bible, the nature of early Israel as a federation of tribes needs to be examined anew: just what do we mean by a tribe?

28. Josephus many centuries later also seems to have made a connection between the Abraham narratives and the Hyksos rule in Egypt (*Against Apion* I. 14–16).

29. Posener, *Princes et pays*, pp. 39–42, is overskeptical.

30. Even then it was a unity gained at the expense of severing relations with many groups, especially the Samaritans—and it did not last long.

THE NATURE OF ANCIENT TRIBES

It may come somewhat as a surprise to many biblical scholars to find a modern anthropological definition of a tribe as: "an association of a much larger number of kinship segments which are each composed of families. They are tied more firmly together than are the bands, which use mostly marriage ties alone." "Thus a tribe is a fragile social body compared to a chiefdom or a state. It is composed of economically self-sufficient *residential* groups which because of the absence of higher authority take unto themselves the private right to protect themselves."[31] As we have seen in the study of "vengeance," this definition does not apply even to individual tribes of ancient Israel.[32] There are cases of course in which such action was taken, but in every example preserved there is clearly a violation of obligation and an attempt made on the part of the supratribal confederation to punish the offense. That the tribes were self-sufficient residential groups is true, with the exception of extratribal punitive actions. (It must be emphasized strongly that tribal organization has nothing to do with nomadism.)

It is at this point that we must take leave of anthropological descriptions of primitive societies to find some other base for the description of a tribe in Palestine in the twelfth century B.C. In the first place, it may be observed that modern Bedouin tribes refer to the European nations as "tribes," using the same word that they use of their own tribal organization, which at least suggests that we are dealing with a social structure which is actually a form of political organization. When we turn to other ancient cultures, we find the etymological origin of the English word "tribe" as a designation of a political organization in early Rome. *Tribus* is thought by some to derive from *tris* 'three'—reflecting the early division of Rome into three (and later four) tribes. Others derive it from *tributum*, the social group which pays tribute to the central treasury as a political unit.[33] In either case it is clearly a political organization with which we are dealing.

Hebrew *ševeṭ* actually reflects the same historical situation: the *ševeṭ* is the baton, "the staff, club or truncheon, especially one serving as a mark of office or authority." It occurs with this meaning as early as the Song of Deborah—the *ševeṭ sōfēr*, the staff of office held by the one "sent," with delegated authority. The Hebrew tribe is then, by extension, in a familiar

---

31. *PSO*, pp. 111, 114. Italics are those of the present writer.

32. The episode of Judg. 19–20 illustrates the tendency, but it was clearly a violation of the norm. The same "private right to protect themselves" is of course a surviving primitive tribal characteristic of a political state, particularly when international tribunals are in existence—and rejected and ignored.

33. *PRE*, s.v. Similarly, old Etruria: "Also nicht Mantua ist gedrittelt, sondern Etrurien bzw. die Eidgenossenschaft" (*ibid.*, col. 2494). Cf. also *ibid.*, s.v. *pagus*.

pattern, that over which the staff of office rules. It is an administrative unit within the federation, though it is most probable that such units corresponded to already existing social groupings which entered the federation as corporate bodies. Such procedures are extremely regular in the transition from one religious context to another: *cuius regio eius religio* did not originate for the first time in sixteenth-century Europe, as the case of the Khazars well illustrates.[34]

Many ancient societies illustrate the social evolution from this sort of tribal organization to a centralized state. We have already cited the case of early Rome. But Athens also was subdivided into *phulē* at an early period,[35] and the Execration texts cite also the "tribes" of Byblos in the nineteenth century B.C.[36] Unfortunately, in none of the cases do we have much information on which to base secure conclusions concerning the make-up (socially) or the functions of these "tribes." The important thing here is to recognize the fact that a tribe in early historical societies has only incidental similarity to what is classified as a tribe by modern anthropology, and is actually a typologically early form of political structure the real nature and function of which we know very little, even in the early Israelite federation. Furthermore, the anthropologist's analysis of a tribal culture in primitive society shows such contrasts to the Israelite tribes that something other than primitive patterns must be appealed to for adequate understanding. We perhaps do not yet have an adequate and well-documented model.

Yet if ancient Israel, Byblos, Athens, and Rome all have political subdivisions termed "tribes" (in translation) at the beginning of their history, it would seem worthy of further investigation. Biblical traditions include further examples (notably from Edom) of the same sort of complex political organization. As at least a preliminary suggestion which may be relevant, we may turn again to ancient Rome.[37] At an early time we find the population distributed (as always in the ancient world) in a host of small agricultural and pastoral villages called *vici*. Somewhere in the geographical vicinity of a cluster of villages in a territory called a *pagus*, is a fortified town called

34. D. M. Dunlop, *A History of the Jewish Khazars* (Princeton, N.J., 1954), chaps. V and VI.

35. *PRE*, 20: 1, cols. 994, 1001. The system seems to be closely related to that of Miletus—an artificial administrative arrangement.

36. Posener, *Princes et pays* E63, p. 94. Note also that the closely related city of Irqata likewise has *whyt* 'tribes' (E61). Cf. Gardiner, *Onomastica*, 2: 205*, where it is derived from *wht* 'village.'

37. *PRE*, s.v. The evaluation is very complex, and likewise bound up with social and eventually political organization. Originally the *pagus* seems to have designated the boundary line within which the peace (*pax* > *pagus*) was kept, but this is a classical example of etymological uncertainty. In other words, we do not know the history of the word or the social history that gave rise to it.

oppidum, which has a complex of relationships to the villages—as a market, as a place of refuge in time of enemy attack, as the location of specialties not available in the villages, and so on. The *pagus* (from which English *pagan* and *peasant* derive) is a natural growth which corresponds to similar residential communities in many parts of the Near East.[38] One thinks particularly of the biblical reference to the town X "and her daughters."[39] It is entirely probable that there was a constantly maintained network of kinship ties between the city populations and those of the villages, through intermarriage, migration from the town to village, and, much more likely, flow in the opposite direction. City populations have always tended to have village origins and to return to the village for wives—and for burial.

But what would happen to this sort of symbiosis between town and village if an external superior power group moved in and assumed control, which seems beyond doubt to have been the case with the movements of the sea-peoples all over the Mediterranean during the fourteenth–twelfth centuries B.C.? (It is worth observing here that even in recent times "imperialists" are always people who arrive by sea, not overland!) The newcomers take control of the *pagus*, and the customary relations between city and village become unpredictable. A Hivvite prince seduces a village maiden; reprisals take place which disrupt the relationships profitable to both sides.[40] The military aristocracy of the city kills the goose that lays the golden eggs, and the majority of the villagers who produce the food supply for the city are alienated and ready for any movement which will give them solidarity on a large geographical basis sufficient to guarantee protection from the superior military technology of the newcomers whose actions are either insensitive or contemptuous toward the needs of villagers whose economy had nearly always been on a subsistence level.

It seems most probable that what actually happened was the formation of large social units of sufficient size to furnish an effective counterweight to the military aristocracies of the city; in turn, the cities formed hasty coalitions of their own to cope with the new threat to political control.[41] Confronta-

38. Compare also the newly discovered text of Salmaneser III from Tell er-Rimah, which concludes with a list of administrative centers each over a group of villages (S. Page, "Joash and Samaria in a New Stela Excavated at Tell al Rimah, Iraq," *VT*, 19 [1969]: 483–84).

39. Num. 21:25, 32; Judg. 11:26; and elsewhere.

40. Gen. 34.

41. Josh. 10. It seems very probable that the so-called Philistine federation may very well have been the successor to this short-lived alliance, since the central hill country seems to have been firmly within the Israelite control and the Hebronite kingship gone (Josh. 12:10). Jerusalem and Shechem evidently were cities which retained their traditional kingship, probably in some sort of symbiotic arrangement with the Israelite villages in the case of the latter.

tion was inevitable, and the urban military aristocracies met with defeat in unorthodox military tactics. The *pagus* was separated from its *vici*, and the result was the disruption of the technological system, to which every archaeological excavation of the Early Iron Age bears witness.[42]

At this stage we may return to the anthropologist's observations: "... it seems remarkable that a tribe remains a tribe. It seems sensible to reaffirm that *external* strife and competition *among* tribes must be the factor that provides the necessity for internal unity."[43] The difficulty with this statement is that it explains too much, however relevant it may be to the situation of the twelfth century B.C. Perhaps it would be a more adequate historical principle if we stated it another way: "To every political action there is, *eventually*, an equal and opposite reaction." It would be a happy case if this were true; unfortunately, the reaction is always opposite, but whether it is equal or not depends upon two factors: (1) the availability of military and technological power; (2) the existence of internal, ethical control.

In the context of early twelfth-century history, it seems incontrovertible that early Israel had virtually none of the first factor and an unparalleled creativity in the second. It is too difficult to avoid the conclusion that the human race, at least in the West, has seemingly always worshipped the first at the expense of the second. At any rate, it is this which makes jobs for the historians who have been unable to make clear to politicians that policy established for a given situation may have side effects for a century to come. The Decalogue proclaimed this three millennia and more ago: "... visiting the iniquities of the fathers upon the sons to the third and fourth generations of those who hate me...."[44] One can only wish that this optimistic point of view could become socially and historically operative, but it can be only if living human beings would accept it. This would mean that any historical event or situation that occurred prior to 1872 is completely irrelevant to the current social situation, and must be relegated to the category of history from which we ought to learn, but do not. For example, the responsibility of Jews corporately for the crucifixion of Jesus, if it was ever justified (which I reject most emphatically) should have been eradicated according to biblical thought by the time of the revolt of Bar Kochba. Similarly, any grievances of the Jewish community against the Roman Empire stemming from the destruction of A.D. 66–70 would have been wiped out by the time of Marcus Aurelius. One shudders to think of the consequences a thousand years or a

42. The degeneration of artifacts of all sorts in the Early Iron Age thus has nothing to do with the supposed "invasion" of barbarian nomads. In all probability the same deterioration of pottery can be seen also in Sicily at this time.

43. *PSO*, p. 114.

44. Exod. 20:5.

hundred from now of the events of the twentieth century if no ethical principle at least adequately analogous to the Old Testament becomes popularly accepted: the Dead Hand of the past must be permitted to die, for after a century there is no corporate responsibility for a criminal act and no basis for further claims by the victims.

FROM TRIBE TO STATE—AND VICE VERSA

What we are describing in the early historical social structure of ancient Rome corresponds, of course, to that which has long been called the city-state in the Near East. Yet the Execration texts demonstrate for us that city-states were not always that: social evolution has not been a one-way street. To judge from the elaborate public works, especially city walls, of the Early Bronze Age, it seems fairly probable that there was centralization of power among the cities of that time, but the period of catastrophe from 2300–2100 B.C. saw a process of devolution: a transfer of power from central authority to local units. To put it another way, the breakdown of empire to smaller social-territorial organizations is the reciprocal process to that which resulted in the empire formation. The term "feudalism" has often been applied to the process of devolution, with results that confuse more often than they clarify, though the European feudalism may very well be one particular illustration of the process taking place. "The devil you know is better than the devil you don't know," has probably been always the reaction of the majority of populations since the beginnings of civilization a mere five thousand years ago. Therefore there is always a potential in any large society for devolution by withdrawal of support from the remote, impersonal, and centralized authority in favor of a local, personal one which is more sensitive to local needs.[45]

Two factors in the history of the Bronze Age are extremely probable influences which would have tended to accelerate the process of devolution. The first is the fact which we have already mentioned, that political control was demonstrably in the hands of a recent, alien, intrusive element, and was therefore certainly based merely upon superior force and social organization (as well as superior military armaments and technology, viz., the "Philistine" monopoly of iron technology during the period of the Judges). The overwhelming preponderance of non-Semitic personal names among the rulers of Palestine and Syria during the Amarna age furnishes all the proof that

---

45. If the central power is too firmly in control, devolution takes place by the formation of ʿApiru groups (see chap. V), which then may become semi-independent states, as in the case of Abdu-Ashirta.

is necessary for this conclusion, though it may well be granted that the non-Semitic element among the political power structures could easily have been several centuries old by the time of the Israelite revolution, and may well have been virtually a constant in the history of Palestine since the pre-pottery Neolithic. By the very nature of its geographical location on the land-bridge between Asia and Africa, Palestine could never have consisted of an isolated "pure" linguistic and racial population.

The second factor is one which needs very strong emphasis, in order to avoid the worst mistake of reading purely modern ideas into the ancient world. Nationalism, like racism, is for all practical purposes a nonexistent operational concept in ancient history. On a large scale, the identification of persons with a vast territorial state in which the political power derives from the governed is a very recent development. The very language of antiquity, which describes persons as the "slaves" of the king—as subjects, not citizens—must be taken seriously. Furthermore, the political power structure typically derives from supernatural, suprahuman sources, from the world of the gods from which the "kingship was lowered from heaven," as the Sumerian king list put it. The establishment of kingship nevertheless could be recognized as a voluntary act of society (well attested in the biblical narrative), yet, once king, Saul's character was something that transcended purely human or social creation: God gave him "another heart,"[46] which in course of time became an "evil spirit." Thus the ancient experience which became rationally elaborated in the legal-political doctrine of the "king's two bodies," from the Roman Empire to nineteenth-century England, correlates perfectly with the modern observation that "the whole is greater than the sum of its parts." Granting that the various units of society recognized and accepted the king as ruler, his designation as such and his authority did not rest merely upon such social acceptance. Nor in the modern territorial state does the individual have freedom of choice whether or not he is subject to the political authority and its legal jurisdiction. All he can do is try to escape that jurisdiction by flight or socially organized resistance. Such a person or group is precisely what is designated ʿApiru in the Late Bronze Age—"stray dogs," or as Nabal says, "Nowadays, there are many slaves who run away from their master" (1 Sam. 25:10), i.e., the King, Saul.

As a test case for the validity of this operational concept of the king's supra–social, divine authority, we may examine the combination of social and ideological phenomena which attended the division of the monarchy at the end of Solomon's reign. On the social-historical level, the proposed policies of Rehoboam could only have been promulgated on the ground that his authority was not derived from the "consent of the governed," even

46. 1 Sam. 10:9; 16:14.

though it was of course always for their benefit (as defined by the king with his superior insight into the divine counsel). On the same level, the proposed policies were absolutely intolerable to the majority of the population, particularly in the north. The result was a mass withdrawal: "What stock have we in David: what stake in the son of Jesse?"[47] But how can a rebellion be legitimized? According to the biblical record, it was legitimized in advance—by a spokesman for Yahweh, Ahijah of Shilo, who represented it as an act of the same deity from whom Rehoboam's authority was derived.[48] It may be observed, of course, that in such circumstances to predict is to cause—but there were two sides to the issue, and Ahijah's prophecy can hardly have been regarded as the *cause* of Rehoboam's incredibly blind and calloused policy, unless it be granted that (upon the advice of the younger generation) it was a deliberate confrontation—which it may very well have been.

Whether or not Rehoboam's policy was a deliberate test of strength, it is clear from the events as well as from the biblical narrative that two radically opposed ideologies were involved in the issue of the religious foundation of kingship. It was not only a political crisis, but also a religious and theological one, which was to be waged for centuries to come, as its roots reached back to the time of Moses. If one forsakes the modern concept of religion as mere cultic ritual (a frequent definition used by the anthropologists) plus doctrinal concepts about the supernatural world, and looks upon religion as operational value convictions which (like all value judgments) cannot ultimately be historically proven or validated, but which nevertheless determine individual and particularly social behavior, then it is possible to see the great depth of the schism at Shechem and its implications for the subsequent history of biblical faith.

Again, as at Baal Peor, mutually incompatible ideologies were clashing, and the only resolution was through fission. If we penetrate behind the façade of the biblical language by examining the record of human and social behavior, it is possible to reconstruct with some degree of reliability the two positions, and as a surplus yield, obtain an insight into the contrast between the ancient biblical and pagan ideologies as well. Since Rehoboam's and the younger generation's position was the simpler and more familiar to modern ideology, we may begin with it.

As Malamat has recently pointed out, the ancestry of Rehoboam's son Abijah, was so slightly Israelite that he might have been "acceptable" to the Nazi regime.[49] There is no reason to believe that Rehoboam would have

47. 1 Kings 12:16.
48. 1 Kings 11:31–33.
49. Malamat, "Comments on E. Leach: 'The Legitimacy of Solomon—Some Structural Aspects of Old Testament History,'" *Archives Européenes de Sociologie* 8 (1967): 167.

had any understanding of, much less commitment to, the old religious tradition of pre-monarchy days, particularly because he was the issue of one of Solomon's innumerable political marriages, with the royal house (presumably, though we are not specifically told) of Rabbath-Ammon—which was merged with that of David upon the conquest of Rabbath-Ammon (2 Sam. 12:26–31), and consequently had no longer any real existence except that legitimized through the offspring of the Ammonite princess. Similarly, the political cult of the "Lady of Ammon" (Rabbath-Ammon) was in all probability actually transferred to Jerusalem.[50] It is this (among many similar cults) which furnished the basis for the condemnation by Ahijah. In keeping with virtually universal ancient oriental custom, political subjugation of an ancient state was also the subordination of its tutelary deity to that of the conqueror, or actually the identification of the two, especially on the part of the subjugated.[51] Thus does political ambition make a shambles (literally, slaughterhouse) of religion—and for this, religion is usually blamed, not politics, which is sacrosanct, at least in the "modern" world. The Rabbath-Ammon, like the Baʾalat Gubla of the Amarna Letters, is the divine symbol of the body politic itself. The identification of the political power structure with the deity is seemingly one of the most ancient religious structures of thought of ancient paganism. Whether the divine symbol is male or female seems to have no significant correlation with other aspects of ancient religious thought. But the identification of god and state is well illustrated from Assur to Athena to Roma, to name only three of the most important, where the name of the state and the name of a deity are linguistically identical. The same, however, is well attested by the incidence of innumerable baals in biblical tradition, many of which are similarly place names. In very like fashion at a much later stage of human history it seems that every city and town with any pretensions to importance had its own tychē (in Greek), or genius (in Latin). These divine symbols of power and prosperity were meaningless apart from the particular sociopolitical segment for which they stand. One is reminded very much of the propaganda of the Nazi "theologians": "If the German race did not exist, God would not exist."

The success of political and military power, however, introduced into the typical city-state (or pagus) of ancient times a twofold disruption which was operative both in the ideology of the conqueror and that of the conquered. In paganism the most important deity was that (or those) which furnished the legitimacy of the political power itself, and it would be politically functional only if it coincided to a high degree with public opinion.

50. 1 Kings 11:33.
51. Thus the Tyrian Melqart was successively identified with Nergal and later with Hercules. After all, the deity was nothing but the metaphysical symbol of the state in the first place.

Unfortunately, we have next to no systematic information concerning the relationship between religious cults and social contrasts—to which modern politics is so extremely sensitive. Do the thirty-five gods of Mari justify the conclusion that there were a similar number of religious communities? Such a conclusion would be naive, but there must be some sort of correlation between such a large number of cultic specialized functions or contrasts and the social structure of this ancient empire. A deity which has no important social following is not likely to be historically operative,[52] and it does not seem probable that the many priesthoods involved would all have specialized functions which were completely independent of social or familiar contrasts within the empire.

The social foundations of ancient polytheism have been subjected to virtually no systematic analysis, but our interest in the subject at present is limited to the relationship between the divine head of the pantheon at any given power center and the subordinate deities of segments within the state or empire. It must be recognized that very small (by modern standards) sociopolitical units had centralized kingship and the correlative political divine symbol in the Bronze Age. The hierarchical structure of the pantheon is a function of the political segments of the population of the state itself, at least in part—it is necessary to admit that factors other than social structure are most likely to have been involved in the pantheon formation at any given location, particularly because of the tendency toward priestly specialization in such fields as healing and law, for which excellent examples exist in early Mesopotamia.

Nevertheless, no state creates its own religion *ex nihilo*. As in the case of Jeroboam at Dan and Bethel, all the state can do is create a prestigious cult center under the ideological control of the king, and proclaim to the population: "behold your god(s)."[53] In the royal cult and in the royal administration they become real. The pre-state functions of the god or gods in the experience of the subject people which have to do especially with economic concerns and with protection from hostile attacks have now been transferred from the realm of the unpredictable unseen world beyond the control of (but not necessarily inaccessible to) humanity, to the realm of the state which is the maximization of human control. It is the divine power now incarnate in the state or even the person of the king which guarantees the success of the daily economic activities of the subjects, just as it is the king who guarantees the military protection with the same divine delegated authority. No one in modern times is likely to quarrel with this doctrine seriously; the large social

52. Note the taunt of Baba to Re-Har-akhti in the contest of Horus with Seth: "Your shrine is empty" (*ANET*, p. 15).
53. 1 Kings 12:28.

organization of the state seems at least to be indispensable to economic organization which can support a high population density. Yet it is necessary to ask whether the centralized territorial monopoly of force really produces anything other than war, law, and taxes, plus a symbol of suprasegmental unity. Above all, it is necessary to ask whether a political state is large enough to be functional in a rapidly rising population and a similarly expanding economy with cross-cultural contacts.

The answer in ancient times was the rise of empire. If the deity of the king ruled over all the earth, then the king should on the historical level do so as well as the "son of god."[54]

This brings about, then, the second disruption or discontinuity of religious tradition. The local political symbols of peoples subjugated by force may be identified with those of the conqueror (in an appropriately subordinate fashion, of course), but as the alien regime becomes increasingly intolerable, the local symbol becomes increasingly important as the rallying point for political resistance. Perhaps the best illustration of this point of view comes from the speech of the Assyrian general in 2 Kings 17, where he cited historical precedent to prove the futility of such religious parochialism. Furthermore, it has often been observed that political rebellion in Old Testament times is almost always accompanied by some sort of religious reform movement—with varying success.

It must be emphasized, however, that the usual pattern by which religion and political issues are identified in ancient paganism is only in part provable for ancient Israel. In the first place, we have very meager evidence that Yahweh, the God of Israel, was ever identified either with the gods of the vanquished[55] (e.g., in the time of David and Solomon), or with the gods of the conquering heroes—of Assyria or Babylon. There is, however, good reason to believe that functions attributed to their gods by the pagan conquerors were in Israel attributed to the workings of Yahweh, the God of the vanquished.[56] The only means by which this could be made plausible enough to deserve preservation in the biblical canon is emphasized strongly enough, and represents the prophetic continuation of that aspect of ancient Israelite religion which alone has no parallel in ancient paganism: the proclamation that political realities are a function of the religious ethic. A nation which conceives of God as merely a symbol of national sovereignty at the

54. Ps. 2:7–8.
55. As argued above, this must have been true of the deity of the Gibeonite high place, and perhaps of the god of the Kenites. It seems certain that tradition has expurgated from the record most of the evidence of the paganizing trends throughout most of the period of the state.
56. Cf. chap. III: the "Vengeance" of Yahweh.

expense of the absolute obligations which that God requires deserves the fate of the pagan nations whose god is overtly merely a social symbol of power— a *baal.*

If the constant problem of a revolutionary movement has been to find a source for its authority that would give it legitimacy and so give its laws a binding validity, it is very much worth considering that the biblical revolution faced a quite different situation and saw the problem as quite other. Hannah Arendt has pointed out that the revolutions of the recent past "presuppose not the breakdown of religious beliefs as such, but certainly their utter loss of relevance in the political realm. . . ." At the same time, ". . . we can hardly avoid the paradoxical fact that it was precisely the revolutions . . . which drove the very 'enlightened' men of the eighteenth century to plead for some religious sanction at the very moment when they were about to emancipate the secular realm fully from the influences of the churches and to separate politics and religion once and for all."[57]

The biblical revolution took place in a different context of human thought and experience. A political organization with a centralized power structure was a fairly new social phenomenon to the populations of Palestine and Syria, and very often it was extraneous in origin and based simply upon superior force. It was the relevance not of religion, but of politics (as known at that time) which was tried and found wanting. As in Greece and Rome,[58] the divine inspiration of the laws was evidently not regarded necessary in Mesopotamia either: it was not the laws, but the *imperium* which the gods created.[59] Even in the Old Testament the earliest law code is strikingly "secular," as many have pointed out in various ways. The concept of law as "the notion of a range or province, within which defined power may be

57. Hannah Arendt, *On Revolution* (New York, 1963), pp. 185, 186.

58. *Ibid.*, p. 186.

59. It is most significant, however, that the original usage of the word *lex* signified "intimate connection or relationship, namely something which connects two things or two partners whom external circumstances have brought together. . . . Only after Aeneas and his warriors had arrived from Troy, and a war had broken out between the invaders and the natives, were 'laws' felt to be necessary. These 'laws' were more than the means to re-establish peace; they were treaties and agreements with which a new alliance, a new unity, was constituted, the unity of two altogether different entities which the war had thrown together and which now entered into a partnership" (Arendt, *On Revolution*, p. 188). It must be clearly emphasized, first, that we are faced here with a precise analogue to the formation of a people, a new unity, by covenant, which is the main thesis of these essays on the origins of ancient Israel. Second, the divine function is presupposed (though not referred to by Miss Arendt) in the making of covenants in the ancient world—the sanctions that guarantee obedience to solemn promises made. The difference between Rome and ancient Israel is the fact that Yahweh is actually a party to the covenant, and therefore the source of the *leges* in the Latin sense.

legitimately exercised,"[60] may well be considered a remnant of primitive tribalism which has been the bane of political reality ever since.[61]

The starting point of politics is the concern for power, but the whole theme of early biblical history—and a recurrent theme throughout—is the rejection of power. To put it in the framework of the Creation narrative, God, who created all, placed the world of subhuman species under the domination of man, but man cannot be thus dominated by man—only by God. Laws are then mutual understandings among human beings, culturally determined and relative—as well as changeable. It was not until the ancient pagan concept of the divine election of the king was brought into biblical thought that law began to be thought of as absolute.

Nevertheless, there were absolutes in biblical thought before the Monarchy, but they were completely contrary to the Greek idea of law. It seems probable that any absolute is so subjectively only—none can be objectively demonstrated in the sense that it becomes a determinant of behavior, except on the part of the one who holds it to be an absolute. From this point of view, it follows that the idea of territory or range (geographically) is entirely irrelevant to early biblical absolutes. Further, the concept of exercise of legitimate power is also irrelevant, for the power, like the absolutes, derives from God alone. Thus we see emergent in the formative period of ancient Israel a new society, and a new ideology which becomes operative in a way quite different from ancient and modern paganism—one which creates enormous difficulties to modern understanding, especially by intellectuals.

The absolutes have nothing to do with power or politics—religion had already been made irrelevant to normal politics by Moses—but with human beings and their ways of behavior. The foundation of the community had nothing to do with a social agreement concerning divine legitimacy of social power structures—this entered from paganism with David and Solomon—but with common assent to a group of norms which stemmed from no social power. If the history of Palestine for two centuries (at least) before Moses had been the sorry record of political attempts to maintain and to expand control, the biblical covenant was a systematic proclamation that no one was in control, and any social organization was a secular business that depended entirely upon its demonstrated value to human beings—and its willingness to remain within the ethical bonds to which all members of the community were obligated.

60. In other words, the territorial and ideological boundary line is constitutive. A true religious ethic in the biblical sense can recognize neither.

61. International private law is still a chaos of insoluble problems.

In other words, the covenant-Decalogue established common norms binding to all members of society—and the society consisted of those who accepted the common norms, the first, foremost, and most revolutionary of which was the rejection of all obligation to "other gods." In practice, this meant the refusal of servitude to those dynasts who claimed to be "chosen" by those *baals*. Thus does David reproach Saul with driving him from the community saying, "Go, serve other gods."[62] The only alternative was death. It was probably not until Solomon that the normal concept of law of the ancient pagan world became politically dominant in ancient biblical life. Conversely, religion became defined increasingly as the formal cultus, especially of the established temple—a thoroughly pagan concept which negated the whole of the community history since Moses.

It was the Mosaic period which constituted revolution; with Solomon the counter-revolution triumphed completely, only to collapse under the same weight of political tyranny and arrogance which had so much to do with the troubles of the pre-Mosaic period. The continuing revolution is represented by the prophetic movement, and culminated in the Christian reformation.[63]

The real issue was a fairly simple one: whether or not the well-being of persons is a function of a social monopoly of force, or the consequence of the operation of ethical norms, which are values determining the behavior of persons in society; whether to put faith in armies and armaments, or in the unpredictable Providence which guarantees the validity of the ethic— though not the reward. What is further at stake is the whole concept of persons: as an ass driven with a stick and seduced by a carrot, the rewards and punishments which imperfect society uses to maintain social control and which are always irrational; or as persons who have transcended the pettiness and meanness of so much of humanity which cannot value itself except by the yardstick of the amount of control a given individual has over the goods, services, and behavior of other persons.[64] In this process, the vast majority of

62. It was the necessity of subjecting himself to the symbols of Philistine absolutism to which David refers; symbols like these were historically real—as was the regime. Such a statement has nothing to do with alleged "henotheism," for the religious statements in the Bible are concerned, not with philosophical metaphysics, but with the choices and symbols of human beings.

63. This is not to deny that rabbinic Pharisaism was similarly a reformation; the culmination, however, *could not* and cannot be identified with *any* ethnocentric paro-chialism.

64. It is this position, already very well illustrated—and condemned—in Jer. 9:23–24, that seems to be virtually the only sort of motivation recognized in modern social science and public news media: the power of economic accumulation, of institu-tionalized position (whether of government, church, university, or dissident student movements—it is the same: based upon the blind, thoughtless subjection of individuals to language which has relatively little connection to objective reality), or of technical knowledge = wisdom.

humanity must be reduced to valuelessness,[65] and the way is prepared for the new revolt—dedicated to greater pettiness and meanness in the majority of cases.

The Sermon on the Mount should have brought discussion to an end, but it eventually began anew. God is not an authority for political manipulation; and any action which is induced by the fear of social punishment or the hope of social reward has nothing to do with religion or God:[66] "They have received their reward." The sanctions of religion are beyond history, and the sanctions of society therefore have nothing to do with religion. Those of the Kingdom of God are not among the perennial masses whose expectations for economic gain have furnished the social power base for the endless series of political tyrannies which make of human history such a tragic farce. Perhaps Moses was, after all, right. Perhaps it is politics, not religion (in the biblical sense) which is irrelevant. It is hardly compatible with any concept of human dignity to say that the art of manipulating the monopoly of force is the god inferior to none in the hierarchy of social values.[67]

65. One wonders whether this is not the best explanation of the feminist revolt movement. The underlying value system seems to be the same as that which the movement condemns—the quite sincerely held nonsense that the only valuable roles or activities are those that bring wealth, public attention, and control over other persons. In the meantime, real needs such as the *giving* of attention, love, and some example of responsible adult behavior is despised.

66. Or, ultimately, with being human: it is merely the ethic of the ass, who is at least useful.

67. It seems increasingly clear on the university campus at least that any sociopolitical order can survive only on the basis of a widespread apathy. This is simply the recognition that there are not very many issues worth fighting about, and those cannot be solved merely by shifting the locus of power from hogs to weasels, or the reduction of orderly process and rational thought to slogan shouting.

VIII

TOWARD A BIOGRAPHY OF GOD:

RELIGION AND POLITICS

AS RECIPROCALS

At the present time, the bankruptcy of much theology is illustrated by the (now defunct) "death of God" movement and the proclamation that the language of religion must become political or at least sociological. It seems appropriate to examine anew the whole question of biblical religion and politics in historical perspective.

It is not only for the theologian that the present situation is uncomfortable, if not intolerable. The problems that arise out of what has been termed "religious pluralism" make life difficult for the judge and the legislator, to say nothing of the man in the street, who experiences the problems but cannot verbalize them effectively. The large-scale withdrawal of the younger generation (which comprises over half of our total population) from traditional religious communities often toward new associations dedicated to the seizure of power is another index of anticipated change in the future. In spite of the undeniable and extensive changes which are in process, it seems to me particularly appropriate to paraphrase the old cry at the death of the king: "The king is dead; long live the king: religion is dead; long live religion." The death of a form is not the death of a function or a value.

For the ancient Near Eastern specialist, the radical distinction between religion and politics which has long been traditional, if somewhat illusory in the West,[1] is virtually impossible. If my historical analysis is correct, it

1. It is illusory, first of all, because religion is thus identified with social organization rather than as an operating complex of conviction, faith, and behavior. Secondly, if we can recognize as religion the complex of state-related myth, cult, and symbol in ancient Canaan, Babylon and Egypt, functional analogies in modern time should also be so classified as religion. The "cult" of Chairman Mao, who "triumphs completely over all objective reality," is strikingly analogous to that of the ancient pharaoh, who also had peculiar talents in the same direction. Among politicians, these two examples are by no means isolated.

was not until the Sinai event that political power structures were secularized systematically, though they were usually disvalued, especially beyond their own borders. It was suggested also in chapter II that during the Iron Age there seems to have been an increasing tendency to identify individual human acts and experiences as manifestations of divine power or revelation, though to draw a radical contrast between Bronze Age and Iron Age cultures in this respect would probably not be tenable.[2]

If we label as "religion" the ideologies and behavior of ancient human beings that are clearly determined by the conviction (or political propaganda!) that the king (who *is* the state) and his political power are the ultimate manifestation of unseen, transcendent factors that determine the future and the total well-being of society, why should there be any hesitation to classify under the same label similar ideologies and behavior patterns of the modern world? To refuse to do so on the ground of sharp contrasts between the ancient and modern modes of expression and communication is to identify religion merely with forms that are historically and culturally conditioned rather than with those complexes of motivation and attitude which find expression in the forms known to us from ancient language and art.

Rather than spend many words on verbal definitions, let us examine the phenomena in action. In a political power structure, the obligations of individuals are a function of status in the social system. Where there is no relationship, there is no obligation. A gruesome illustration is furnished by the Mari letters. The king wrote to Yasmakh-Addu to this effect: that since a peace treaty with a certain tribe had not been ratified, the official should take the hostages from the group who were in his custody, bring them out by night, dig their graves, put them to death, and bury them.[3] The same motif is operative when labor unions proclaim, "no contract (i.e., no covenant), no work." Where there is no relationship pre-ordered, frequently if not always by covenant, there is no obligation, and consequently there is no security of persons or property. It is quite interesting, therefore, to find that a modern jurist has argued that citizenship consists at least partly of an "implied covenant," which he appeals to as a basis for the citizen's obligation to obey laws which he may have had no real part in establishing—in fact, he may have voted against the representatives who voted for the laws.[4] This is

2. Any superficial reading of the epic poetry of Homer or Vergil gives abundant illustration of the identification of human persons or acts as such manifestations. The difference between this and the Bible needs further consideration. As a preliminary observation, cursory reading of Homer and Vergil gives the strong impression that such theophanies or "godlike" activities center around situations in which a human being exercises control over others, either in war or in political debate. This was hardly a temptation to the prophets.

3. *ARM*(T) I, 8:5–17.

4. See J. Tussman, *Obligation and the Body Politic* (New York, 1960), pp. 7ff.

doubly interesting as an illustration of a very common motif in contemporary thought: the proposition that the political state must rest upon something other than brute force. One jurist even registered dissatisfaction with the fact that continental philosophers of law could write thousand-page monographs on jurisprudence without ever once mentioning or even admitting that law is the orderly means by which society exercises its monopoly of force, and that upon force the law ultimately must rest.[5] A law not enforced is simply no law.

Since every civilized state is a monopoly of force, illustrated by the fact that no private citizen can with impunity use force against another except in immediate self-defense, must it follow, as it has throughout most of human history, that there is no representative of the gods quite so adequate

5. Since there is an enormous literature on the subject it need not be further discussed here. However, in view of the great confusion in recent literature over law and covenant in biblical thought, it may be worthwhile to summarize the contrast between them, since social force is one of the many criteria that marks that contrast to which all ancient societies seem to have been sensitive at least at some time in their history.

	COVENANT	LAW
Purpose:	To create new relationships in accordance with stipulations given in advance. (Cf. marriage.)	To regulate existing relationships by orderly means.
Sanction:	Punishment and reward meted out by powers not under human control, namely, God or the gods: curses and blessings.	Punishment defined by and administered by social organs.
Norms:	Undefined except by the fear of God (= conscience) of the actor.	Defined by society, and necessarily in formal terms.
Binding:	By voluntary act of individual, usually ritual in nature.	By enactment of legitimate social power, regardless of individual's attitude.
Validity:	Unlimited, since it is bound to the individual wherever he may be.	Territorially bound—no validity beyond the boundary.
Time reference:	Future—it is a solemn promise as to future behavior.	Past—it does not operate as a social reality until a violation has taken place; it has a future orientation only in that it is a prediction of what courts will do in the case of a past violation, and therefore may be a deterrent to proscribed forms of behavior.
Procedure:	Recognized in the dynamics of historical process.	Ritual form prescribed in advance by society.

For further discussion, see the article "Covenant" in a forthcoming edition of *Encyclopaedia Britannica*.

as the omnipotent monopoly of power? This proposition could not be and was not effectively challenged so long as the gods were thought to be primarily invisible force structures which delegated power to human agencies, on the one hand, and determined events in the natural world which led to good crops, fertility of flocks, but also to epidemic disease, drought, and crop failure, on the other. The only question which could arise was the perennial question of which of the competing power structures is the one really chosen by the gods. This was and still is usually answered by history: whatever is, or at least continues to be within the experience of the one called upon to make the choice, is right. Consequently, even by the time of the Sumerian King List, the history of political domination is merely a theological affirmation of what actually happened.

The Platonism involved in the old Sumerian affirmation, "when the kingship was lowered from heaven, the kingship was in Eridu," entailed the necessity to recognize the political legitimacy of past regimes in order to affirm the legitimacy and divine nature of that which was contemporary from the point of view of the scribe and his social context. The least one can say is that this concept of the kingship, i.e., as the supreme power, transferred from one city to another from the Persian Gulf to Mari, is at least adaptable to new circumstances, which is a favorite modern definition of intelligence.

If a Supreme Court justice can with good conscience state as he did quite recently that so far as human history is concerned, justice[6] is an irrational ideal, then we should have fairly good grounds for concluding that no monopoly of force can possibly be regarded as the ultimate arbiter of values. So far as I can see now, we have no written document which can reasonably be interpreted as embodying this point of view prior to the Sinai covenant, though I strongly suspect that many of the nonurban populations of the ancient world would have agreed with very considerable enthusiasm.[7] The ancient pagan deification of force is of course a way of stating that the primary function of any political system is to secure and maintain its own power, since a "house divided against itself cannot stand." All that any legal system in normal human society can expect to do is to secure its own continuity by maintaining a delicate balance beween conflicting interests, like a university dean.

One intention in this argument is to show that it is not only pious verbiage on our coins which states "in God we trust," but it is recognized

6. Cf. Cardozo, *Selected Writings*, p. 224: ". . . it remains to some extent . . . the synonym of an aspiration, a mood of exaltation, a yearning for what is fine or high." The lecture was delivered in 1923.

7. E.g., the Benjaminite's retort to the administrator in the Mari letters, *ARM*(T), III: 38.

by statements made by the highest courts of our land that the monopoly of force has a limited validity and function. This is much more effectively put by Roscoe Pound in his *Introduction to the Philosophy of Law,* when he explains why in fact the law has only a limited ability to protect persons. As we have it now, it was developed over the centuries primarily to secure persons in the enjoyment of property; furthermore, there is an infinity of social organizations other than the law in our society, and their function in society, at least in part, ought to be to guarantee the security of persons.[8] In another way the same is true even in regard to the punishment of crime; there are a number of legal decisions in the nineteenth century which state that in a particular case the law is for technical reasons not able to carry out the punishment of persons, and that therefore they can only be handed over to a Higher Court.[9] So far as organized society is concerned, they must be set free. The exercise of force is a limited and imperfect instrument.[10]

It would take only a minimum of historical imagination to see that in early societies characterized by a minimum of sociolegal organization, such an appeal to value systems rather than technical formal procedures would probably have been highly developed. This I believe was actually the case, and for this reason the ancient world even in the Late Bronze Age before Moses had a sophistication which some are now beginning to regard as an ideal for contemporary society. The pre-Mosaic political paganism is the program which Harvey Cox is recommending for the twentieth century,

8. Pound, *Philosophy of Law,* p. 104: "Not all of social control can be achieved through the legal order. . . . Relief from the burden of inequality of economic condition, relief from want, relief from fear, insurance against frustration where men's ambition outruns their powers are laudable humanitarian ideals. But, although many things men had long felt were impossible have come to pass in our time, one may well feel that much, at least, of the laudable humanitarian program is beyond practical attainment by law." And also: "It may be that some part of what is sought will prove best left to nonpolitical agencies of social control" (*ibid.*).

9. "The influences of human conduct, good or bad, are far-reaching, and are often seen and felt in consequences exceedingly remote, but uncertain and complicated. It is simply impossible that municipal law should take cognizance of all these consequences. From necessity, a large share of them must be left to the jurisdiction of public opinion, individual conscience and, finally, to the retribution of another world" (M. Plant, *Cases on the Law of Torts* (Indianapolis, 1953), p. 568 (from *Wells vs. Cook,* Supreme Court of Ohio, 1865).

10. Note the comment of Cardozo on the role of fear in the development and exercise of law: "I do not know whether it can ever be different, or so at least I say in hours of depression. Perhaps this is what law means. It is the medium, the instrument, by which society represses conduct which awakens fear of such intensity as to make tolerance impossible. We shall rationalize law only when we rationalize our fears—our fears and our wishes, the counterpart of our fears" (*Selected Writings,* p. 94). The statement was made in the relatively tranquil time of the year 1928.

since God is dead.[11] The reduction of religion to politics merely signifies a
retreat to the Bronze Age, not becoming modern, for the simple reason that
a customary law rests ultimately upon the grass roots of customary acts of
the majority of reasonable citizens. It is the infinity of spontaneous acts of
persons based usually upon presupposed but nonverbalized values and
interests which creates the models that the law enforces. It would be the
height of injustice if the law were to enforce someone's ideal value system.
Summum ius, summa iniuria.

So far as most of the classical religions are concerned, after the creation
of the community itself, the second major social function which they have
usually performed to a greater or lesser extent is to furnish the motivation at
the popular level of custom for transforming a verbalized value system into
behavior and behavior patterns, and subsequently to perpetuate and maintain
that value system.

It is of course quite evident that such macrohistorical generalizations
must be made with a very considerable diffidence, yet it is at least an extraordinarily productive way of viewing the problem. A hypothetical reconstruction of the history of religions in broad outline would go as follows:
in a social field which is beginning to break down and become unsatisfactory
to large numbers of individuals, a new message of a new religion is presented.
So far as the formal symbols and ideas are concerned, they must be adaptations of already existing concepts and forms if they are to communicate
anything at all to those to whom the message is presented. The presentation
is of the familiar, but it is transferred to something new. The covenant
structure at Sinai is of course a classical example, but by the simple process of
transference from the realm of politics and power structures to the realm of
religion of personal relationships and ethic, it became something radically
different and functioned in a radically different way. This is another illustration of the law of functional shift. Most academic creativity is based upon
such transference and sometimes an unexpected functional shift. A famous
case is the cloud chamber, so useful in nuclear physics, which was inspired by
observing the bubbles rising in a glass of beer.

If it be thus true that the formative period of a new religion characteristically produces only a bare minimum of really new forms and concepts, it
follows that for this period at least, forms are chosen because they are useful,

11. *Secular City*, p. 254. The idea that politics brings unity and meaning to human
life and thought would apply beautifully to the royal metaphysics of ancient Egypt, but
one wonders whether Cox has ever had any exposure to political movements among
university undergraduates. Politics now is just as divisive as denominationalism was in
the nineteenth century. The statement is a yearning for a new Constantinian establishment, for which every theologian wants to have preeminence in drawing up the new
Nicene Creed—very up to date, of course.

because they serve a prior concern. Forms themselves are not an end, but a means, and most of them have a prehistory in the previous history of the social field within which the new community arises. It is frequently the case, however, that in order to protect the contrast between the old and the new, the new community chooses particular forms which are already archaic and perhaps living on not in social life but merely in memory, or in the more remote backwaters of society where they have been preserved. They may be borrowed from outside the social field, but in every case there is a radical shift in the meaning, the associations, and the relationships to reality. The most important new creation of a new religion is the new community itself, which must of course appeal most strongly to those who have no community or who are induced to desire something better.

Within an incredibly short time the new community develops a system of communication in ideology, ritual, ethic, and social organization, and to the founder eventually is attributed everything which the community deems important to its own life.[12] Almost always conflict situations arise, for nothing is more likely to arouse conflict and even violence than the process of large-scale withdrawal of persons from one community allegiance to another. The consequent growth in size and complexity of the new community results in the actual acquisition of social power, and in every case of the historically originated religions, the new religion becomes identified with a political power structure within three centuries, usually in much less time. In the case of Zoroastrianism and Christianity, it became established by an existing political system. In the case of early Israel and Islam, it developed its own upon the wreckage of an earlier one.

It seems quite clear that no political state creates its own religion, but something analogous to religion is necessary to the state if only because there must be an allegiance on the part of most citizens which will make it possible for the state to carry out its functions in manipulating its monopoly of force. The state has life and death powers over its citizens both in law and war; it withdraws considerable amounts of private property for the "public good" in the form of taxation. It seems the most natural process then that the religion looking for means by which to guarantee the translation of its value system into behavior becomes allied with a power structure which is looking for an ideology and a value system that will supposedly guarantee the allegiance of the majority of citizens and give it the legitimacy to maintain continuity.

---

12. Thus the familiar attribution of the "oral law" to Moses was a social necessity in order to avoid the accusation of arbitrariness and disruption of continuity. The attribution of all sorts of "pseudepigrapha" to ancient patriarchs should probably be regarded as a similar phenomenon of earlier type.

In the ancient Near East, to return to some sort of reality in this theoretical framework, it seems clear that the state religion is a composite derived from the pre-existent local community religions which were absorbed into the power system. Though the ancient pantheons are of course extremely complex, this process of absorption must have been at least one aspect of the construction of a pantheon. Yet it can hardly have been the only one, for the kingship itself had at its disposal a large bureaucracy of various kinds of specialists, and speculative thought certainly created much in the myths that can hardly have derived from the spontaneous creations of local city and village religions. It is difficult to say much more about the state construction of a pantheon because we have extremely little knowledge about the social functions and relationships of ancient temples and even less about village cults. Little systematic investigation has even been done in this area.

It would seem to be a fruitful and adequate thesis that the pantheon of ancient states is the most normal and easiest response to religious pluralism, and is actually an extension of the process which had already been in existence long before the centralized power structure of the large state existed. In addition, the fertility cult required more than one deity for any adequate rationalization of the deification of the process of production. There is no doubt that the economic productivity upon which prosperity and continuity depended was viewed as a sexual process.[13]

If the fertility cult was the basic type of religious structure in the time before the state, it is only to be expected that the fertility functions of the gods would survive after they have been taken into the state pantheon.[14] Furthermore, as another illustration of the political readaptation of pre-political religion, the function of fertility (and therefore prosperity) tends to become closely associated with the person of the king himself. As late as Iranian religion, for example, the *hvar^ena* of the king, which is astoundingly analogous to the Assyro-Babylonian *melammū*, has such close associations with fertility and productivity that one scholar several decades ago attempted to prove that *hvar^ena* actually meant something like "bountiful blessing."[15] The proclamation of the king or state that the economic well-being of the citizenry was a function of the king rests upon an insight which is not to be lightly dismissed so far as the city population was concerned, but this broke

13. Rib-Addi of Byblos thus complains repeatedly to the king of Egypt that his fields are like a woman without a husband, and that therefore he has no food supply (*EA* 74:17–19).

14. T. Jacobsen, in *IAAM*, p. 200.

15. I owe this idea to my colleague, Professor Gernot Windfur. For a discussion of the term, see Richard N. Frye, *The Heritage of Persia* (Cleveland, 1963), pp. 40 and 252 (n. 49). Note that in Middle Persian the word is written with the Semitic ideogram *GDH* = Gad, 'good luck.'

down where villages produced food on rainfall-dependent farms. For such villagers, the king was not the producer, but merely the consumer.

The purpose of this discussion in part has been to show the various kinds of functions in society which were inseparable from that which we call religion in the ancient world, and further to show the constant relationship between the secular and the sacred as we now conceive of them. Until the Sinai event, and to a large extent even afterward, the only operating distinction between the religious and the secular seems to be the distinction between what was important and what was unimportant. To be sure, the ancient's ideas of what was important seem at first glance somewhat ludicrous. For example, an obscure Egyptian god whose name means something like "hair of the dawn," is probably the royal beard, and the god's priest is the royal barber.[16] Yet after all, everything about the king was important, and the excessive attention given by the public to insignificant details about the life and style of political persons is by no means a trait peculiar to ancient peoples.

The roots of the church-state problem thus strike deep into remote antiquity, and it seems probable that there will never be any lasting permanent solution, nor is such to be expected. The process of adaptation in the early religious community which called itself Yisraʾel is now a classical example closely paralleled by what happened to Zoroaster's message. As many scholars have seen and long argued now, early Israel was the kingdom of Yahweh,[17] or as I prefer to put it, the dominion or Imperium of Yahweh, since early Israel evidently rejected the title "King" for Yahweh.[18] As any king must do, the divine king delegated certain operations of the imperium to persons. We have already seen that even in quite recent English law the king as the source of all justice was present in every courtroom, even though he himself personally never judged a case.[19] As in many a modern church organization, the decisions of local authorities or officials were determined by what they knew to be the policies of the king,[20] and the early laws of the

16. A. M. Blackman, "The Stele of Nebipusenwosret," *JEA* 21 (1935): 4 (n.2). As an illustration of the deification of the Important, see O. Weinreich, *Zwölfgötter-Reliefs*, pp. 18–19.

17. Bright, *Kingdom of God*; G. E. Wright, *The Rule of God* (Garden City, 1960).

18. But cf. Exod. 15:18.

19. Kantorowicz, *King's Two Bodies*, p. 5, quoting Blackstone's *Commentaries*, p. 1270.

20. Compare Cardozo's statement that "We shall unite in viewing as law that body of principle and dogma which with a reasonable measure of probability may be predicted as the basis for judgment in pending or in future controversies." From the concept of Yahweh as sovereign and court of last appeal arises Old Testament prophecy: if human courts reject those principles and dogmas which constitute justice, the Supreme Court may be relied upon to overrule—and overturn. Cf. Exod. 22:21–24.

Pentateuch give excellent illustration of what that policy was and how it operated in concrete cases.

Though it is not a very comfortable procedure to reduce to abstract principles the value system presupposed in the early laws of the Pentateuch, since we have too little information about the actual social and historical context out of which they arose, two propositions seem to be quite valid. The first has to do with that which the modern jurist states is very difficult for the modern law to carry out, namely, guaranteeing the security of persons as opposed to the security of property. In early biblical law, the security of persons is first of all a function of the covenant relationship, though it would be difficult to find it specifically so stated. This is implied in the concept of a holy people, for holiness is always most closely connected with the concept of the property of God.[21] In practice, then, it meant that the value of a person was not a function of his particular role in society. It furnished everyone with a basis for self-respect, a self-valuation which seems to be necessary for personal freedom and integrity. Most closely related to this is the nature of obligation, which is not merely a function of status within a political community, but an obligation permanently binding upon the person no matter what his immediate social context may be, and no matter with whom he is dealing.[22]

To put it more simply, the obligation of a person continued and had to be respected, even if he were in a social context which made it impossible for society to enforce its legal rules of behavior, for example, the major problem of modern states when armed forces are stationed or visiting in foreign lands. For this reason alone, it is not really adequate to label as law the statements of normative behavior we have in the Pentateuch, for law in the strict sense of the term is not law if it is unenforceable—law is essentially procedure—but the obligations were tied to the person, not to the enforcement procedure of society. It is a rather curious and intriguing suggestion, therefore, that the biblical thought really does not include any concept of rights (which is actually quite modern),[23] and any legal procedure is based not upon a violation of rights, but upon a violation of obligation on the part of the defendant. The modern legal doctrine of negligence is a bare survival.

21. Jer. 2:3–4.

22. This is the original meaning of the concept that is later elaborated as the doctrine of the "omnipresence" of God. Again, it is a transference from the realm of politics to that of religious conviction and ethic. Cf. *EA* 264:15–19: "If we ascend into the heavens, or descend into the underworld, our heads are in your hands [= power]." This is the background of Amos 9:2 and Ps. 139:7–10.

23. The elaboration of declarations of human rights belongs only to the nineteenth century and subsequent periods.

It seems clear then that the concept or experience of the Imperium of Yahweh in early Israel already had to transcend the normal ancient or modern concept of the state and its political functions. A community based upon voluntary acceptance of a value system is an ideal which has often been realized in part in human societies, but the voluntary choice of human beings includes a potential range of ethical behavior which is almost infinitely greater than that which can be enforced by the formal procedures of courts of law.

We see the conflict between these two kinds of ethical systems with the beginning of the monarchy. It is unfortunate that the conflict is usually viewed within the framework of modern generalizations about conservatism and progressivism, for the conflict is really between two different kinds of community, the face-to-face personal relationships based upon a common value system[24] and the impersonal power structure of the state which can hardly do more than maintain a delicate balance between conflicting interests.

It is quite clear from our narratives that the king was expected to uphold and enforce the religious value system of the old religious community—the "law of God"—but the attempt broke down under the impact of the ancient pluralism. King David had by conquest incorporated the surviving Canaanite city-states into the state, and the social and religious contrast between the old Canaanite religion and the new Yahwistic religion made it impossible for the state to enforce such laws as that prohibiting lending money at interest.[25] This is by no means a peculiarly ancient problem; in times of rapid social and cultural change there is always extreme stress placed upon the law—and upon religious value systems. As Cardozo put it, when faced with a new situation, it is most tempting to attempt to maintain continuity merely by refusing to change the forms or formulas of the law. To think that this is continuity, however, is sheer illusion. The similarity is verbal only; it no longer has the same relationship to reality—and cannot have the same function in society.[26] Similarly, nothing is gained by merely changing the form if the new form has no functional relationship to the value system.[27]

Just as the state broke down the old tribal structure of early Israelite society,[28] the religiously centered value system broke down and even made

24. Yet including a population of perhaps a quarter of a million people. It was what may be termed an extended and transferable community.

25. From this time on, the dualistic ethic of Deut. 15:1–3 and 23:19–20 may easily have arisen at any time on the basis of custom.

26. "Paradoxes of Legal Science," in *Selected Writings*, p. 257.

27. "Mere change without conservation is a passage from nothing to nothing. . . . Mere conservation without change cannot conserve" (A. N. Whitehead, *Science and the Modern World* [New York, 1948], p. 201).

28. All the enormous speculation about the "lost ten tribes" of ancient Israel is fatuous in view of the fact that states characteristically make earlier tribal structures

possible the sort of corruption of law courts which the prophets describe in the most bitter language.[29] In this situation, the old traditions of Sinai are appealed to, though in this particular context we are interested in the ᶜānān of Yahweh. In the pre-Exilic prophets the coming ᶜānān of Yahweh is remarkably analogous to the storm which is used by the Assyrian kings as the metaphor to describe their attack.[30] At the same time, the later pre-Exilic prophets seem to me quite clearly to illustrate a maturity of insight which is most interesting. It is not the king who is a primary object of attack, but what I should term the inadequate value system of the population, the citizenry, or at least the dominant group which determined public policy.[31] It is highly irrational to blame the state for failing to maintain the delicate balance between conflicting interests when those interests are themselves irreconcilable not only with each other but also with those of the deity from whom the authority of the state was supposed to derive.

It is *after* the destruction of state and temple that the old traditions received powerful reinforcement and again were readapted. It is most tempting now to see the Yahweh speeches of Job within the framework of a theophany quite functionally analogous to Sinai. First, Yahweh speaks from the whirlwind,[32] and I have long been convinced that the Wanderings traditions about the pillar of ᶜānān represent an early identification of the ᶜānān of Yahweh with the desert whirlwinds (which the modern Bedu still identify with *jinn*). The anonymous author of this masterpiece is actually presenting in a most sophisticated way his discovery that the obligation to Yahweh is not merely a function of the reward and punishment system of a social or economic structure now destroyed, but has its own independent origin and validity. The religious as a determinant of behavior exists in an imperfect (because they do meet and coincide) parallel to the sanctions of organized society, and it is irrational to expect a perfect correlation.

To return to an earlier motif, we see then a powerful resurgence of the idea that the manifestations of God become real in the experience and subsequent behavior of human beings. At the same time, far to the East, the Iranian *hvarᵉna*, with its associations with the Babylonian and Assyrian *melammū*, is being represented as the Lord of Wisdom, Ahura Mazda, in the

---

superfluous and functionless. They may survive for a time as mere genealogical status symbols for individuals, as in ancient Rome.

29. E.g., Amos 5:10–15.

30. Jer. 4:13; Ezek. 1:4; Nah. 1:3; Ezek. 30:3; cf. Ezek. 38:9, 16. Cf. also *ARAB*: "Over the whole of his wide land I swept like a hurricane" (Sennacherib, 2:118); "... like the onset of a mighty storm I tore up by their roots" (Esarhaddon, 2:210).

31. Amos 3:1–11; 4:1–3. The women seem to be treated with democratic parity in the latter citation, with which cf. Isa. 3:16–4:1.

32. Job 38:1.

Fig. 21 The Royal seal of Darius, king of Persia. Ahura Mazda flies above the king in his chariot, but unlike the Assyrian representations the deity is armed only with the symbol of authority, not with weapons. British Museum No. 89132. Photograph courtesy of British Museum.

winged sun disk.[33] The seven spirits of Zoroastrianism are mostly abstractions which have to do with the experience and behavior of human beings, just as the seven spirits in chapter 11 of Isaiah are inseparable from personal characteristics.[34]

In later times, we find in the eschatological visions of the Book of Daniel a passage which now seems to be extraordinarily archaic (7:13–14):

> I saw in the night visions,
> and behold, with the ᶜanānê of heaven
> there came one like a son of man,
> and he came to the Ancient of Days
> and was presented before him.
>
> And to him was given dominion
> and glory and kingdom,

33. See fig. 21.
34. Gershevitch, *Avestan Hymn*, p. 10. Each of these "Incremental Immortals" is, however, in charge of an earthly element.

that all peoples, nations, and languages
should serve him;
his dominion is an everlasting dominion,
which shall not pass away,
and his kingship one
that shall not be destroyed.

It is not necessary to comment at length upon this passage, but it is difficult to avoid the conclusion that there is a very deliberate and pregnant contrast intended between the dominion of the one "like a son of man," and the dominion of the four great beasts, "kings who arise from the earth," particularly in view of the monsters who are such constant attendants of the great winged sun disk and the ancient kings. Since we are not dealing with clear and precise statements of political philosophy, but with value judgments and expectations of something better, it would be futile to think that any precision of interpretation can be proven. Very diffidently, I would suggest that the whole issue of iconography is involved here. The prohibition of images correlates with the fact that man himself is the image of God, according to the Creation story of Genesis 1. Furthermore, I believe the same story intends to state that the delegation to man of authority and dominion over the natural world does not include any similar dominion over man himself.[35] In theological language, the political power structures are not an order of creation or an image of God.

Nevertheless, the old ideal of the Kingdom of God was very much alive and represents the heart of the teachings of Jesus. We have seen already the source of later church theology about Christ, but its origins in the Sinai tradition are now absolutely certain from the narrative in chapter 17 of Matthew:

And after six days, Jesus took with him Peter and James and John his brother, and led them up a high mountain apart. And he was transfigured before them, and his face shone like the sun, and his garments became white as light. And behold, there appeared to them Moses and Elijah, talking with him. And Peter said to Jesus, "Lord, it is well that we are here; if you wish, I will make three booths here, one for you and one for Moses and one for Elijah." He was still speaking, when lo, a bright cloud overshadowed them, and a voice from the cloud said, "This is my beloved Son, with whom I am well pleased; heed him." When the disciples heard this, they fell on their faces, and were filled with awe.

35. The same point is made by omission in the Creation story of Gen. 1, which contrasts sharply to the major concern of the Babylonian Creation Epic, namely, that creation is the ground of being of the Babylonian state (or king), to which is delegated dominion over all that the state had succeeded in subjugating by military force. All outside is the realm of Tiamat—the powers of chaos and evil. Dualistic mythology is always essentially political. Compare also the "Manual of Discipline" of the Qumran scrolls, and the Book of Revelation.

In virtually every occurrence of the word ᶜānān in biblical Hebrew, the Septuagint translates with the same word we have in this passage: nefelē. In addition, the bright cloud is said to "overshadow" (episkiazein) them, which is again the Septuagint translation of Hebrew šākan,[36] the verb used in the passage which describes the transference of the ᶜānān of Yahweh from Mount Sinai to the tabernacle (Exod. 40:35). After the construction of the tabernacle according to the specifications of Yahweh at Sinai itself, according to P, a motif familiar from pre-Mosaic times,[37] Moses could not enter "for the cloud covered (episkiazen) the tent of meeting, and the glory of the Lord filled the tabernacle." It is clear that this narrative in P is modeled after the story of the dedication in 1 Kings 8 of the Temple of Solomon, where the priests also "could not stand to minister because of the cloud (ᶜānān)." In this passage, however, the cloud can reasonably be identified with the actual smoke from the altar upon which "so many sheep and oxen [had been sacrificed] that they could not be counted." We have already seen that the malʾāk of Yahweh ascended to heaven in the smoke and flame of the altar in the story of Manoaḥ. The conclusion that the smoke and flame of the altar were identified with the manifestation of Yahweh seems virtually inescapable when we compare Leviticus 16:2 with 16:13:

Tell Aaron your brother not to come at all times into the holy place within the veil, before the mercy seat which is upon the ark, lest he die; for I will appear in the cloud (ᶜānān) upon the mercy seat. (16:2)

And he shall take a censer full of coals of fire from the altar before Yahweh, and two handfuls of sweet incense beaten small; and he shall bring it within the veil and put the incense on the fire before the Lord, that the cloud of the incense may cover the mercy seat which is upon the ᶜēdūt, lest he die. (16:13)

This identification of Yahweh with the ritual cloud of incense of course emphasizes that it masks but reveals the presence of Yahweh over the ᶜēdūt, which is itself the covenant, as W. F. Albright pointed out long ago,[38] and this conclusion is reinforced further by the fact that in this ritual the blood of the bull is sprinkled seven times before the mercy seat. This can only be interpreted as a ritual covenant renewal based upon the old tradition of chapter 24 of Exodus, in which half of the blood was thrown against the

36. For the extremely ancient and persistent religious significance of this root, see F. M. Cross, "The Priestly Tabernacle," BA 10 (1947): 65–68; reprinted in BAR 1: 201–28. Allowance must be made for the overemphasis on the influence of nomadism as a base for explanation.

37. King Gudea of Lagash was similarly given blueprints for the temple he was to build. See Gudea Cyl. A V–VII, in A. Falkenstein and W. von Soden, Sumerische und akkadische Hymnen und Gebete (Zurich, 1953), pp. 142–44.

38. The paper is as yet unpublished. See now Hillers, Covenant, p. 161, n. 4.

altar, and after the reading of the covenant text and the response of the people: "All that the Lord has spoken we will do, and we will be obedient," Moses then took the other half of the blood and threw it upon the people saying: "Behold the blood of the covenant which the Lord has made with you in accordance with all these words."

In the temple cultus, however, according to the text of chapter 16 of Leviticus, where we should expect the binding of the people to the obligations of the covenant, instead there is a further ritual which removes the guilt of the people of Israel in the form of the scapegoat. We have here a most interesting readaptation of the old Sinai covenant, which may possibly throw further light upon the alienation of the prophetic tradition from the mainstream of the religious tradition associated with the Jerusalem cultus, which reached its climax in Jeremiah's Temple Address, for which he narrowly escaped execution for blasphemy and treason.[39]

After this long excursus, we return to the last reformation of the traditions of Israel included in the biblical canon: based upon the fact that the ʿānān of Yahweh hovered over the tabernacle, Peter wished to build three tabernacles for the three manifestations of Yahweh—Moses, Elijah, and Jesus. Peter did not understand that the manifestation of God was intended as an overshadowing of *them* personally, the motif found already in Solomon's speech of dedication in negative form ("not made with hands"),[40] and in the speech of Stephen, reiterated. The sad story concludes with Stephen's vision of the Son of Man and the glory of God.

We have seen in Egyptian narratives a century or two before the beginning of the Christian era that Horus was identified, recognized, in manifestations of power: in war, in nature, in conflict. We have seen the age-old continuity of this tradition in the extrabiblical sources, but also what we believe to be an increasing tendency to recognize manifestations of deity not in power exhibitions, but in qualities of personality. In the last reformation (that of the new covenant), the focusing of this tendency is identified in the covenant obligation of love, derived from Deuteronomy and Leviticus.[41]

This main theme of the Hebrew Bible as well as the Christian scriptures seems to be a major source of embarrassment to modern man. On the basis of the popular mass media, it is rather tempting to conclude that love in contemporary culture seems to be identified only with the immediate cause of the population explosion.[42] Since the biblical thought seems to be the

39. Jer. 7.
40. 1 Kings 8:27; Acts 7:48; and the quotation from Isa. 66:1-2.
41. Deut. 6:4-5; Lev. 19:18. Cf. Matt. 22:35-40, and John 13:34-35.
42. It is a curious fact that the word is also becoming a slogan for dissident movements in modern society. It remains to be seen whether the ultimate objectives are personal relationships or political power.

reverse of the Platonic/mythological, it would seem to follow that in biblical thought love is only a word with no reference to reality apart from those manifestations in human experience and motivation—things that happen to us on the one hand, and overt actions, behavior of human beings on the other, which are recognized as worthy of the word label. It is clear that in biblical usage love is first of all a label for the fact that persons have established and continue to maintain personal relationships with others, in which the concern for the well-being of the other is recognized as an obligation that takes precedence over other concerns such as the exercise of power or profiting at the other's expense.

To return to our main theme of religion and politics as reciprocals, it is important to recognize that love is the basis of any state, as we can see in ancient literature from the letters of Rib-Addi to Deuteronomy and Shakespeare. When the scoundrel King Edward asked the widow of his victim whether she loved him, she answered that it was the duty of every subject to love the king.[43] As in modern civilization, that was not where King Edward's interests lay. We are politically and economically determined to a dangerous extent. The consequent real hazards to persons and to communities based upon those real actions of real persons which the Bible calls love, makes it at least possible now to understand the emergence of new religious communities in the past and the simultaneous reorientation of the value systems which determined behavior. We need careful examination of the functioning value systems of ancient societies as well as our own. In spite of the vast gap in the system of communication, the vast differences in the uses of words between ancient and modern man, the hypothesis would seem worthy of serious consideration that the established religion of today is not Judaism, Catholicism, or Protestantism, but the deification of power and the deification of the process of economic production. The political, they say, is the art of the possible; but it is the religious that determines *what* is possible.

43. III *King Henry VI*, act 3, sc. 2.

IX

THE TENTH GENERATION:

A SUMMING UP

Some of my friends in the history department at the University of Michigan have long divided historians into two groups. The one is called the stamp-collectors' school of history; they conceive of their task as merely to find and prove some individual historical fact as a stamp collector collects stamps. He who has the most facts in his album, neatly classified and arranged, is the champion historian.[1] The other school is not so easy to label. It sees history rather as the reconstruction of the life and experience of real human beings of the past. Though we need facts, it goes without saying, a fact is not history until it is put into context—until it has received a context, not in the collector's stamp album, but in the complex of relationships of life and community of human beings of a particular time and place. A historical fact is meaningless without its context. For such historians, then, the real task is one of synthesis. One might compare it to the task of the paleontologist, who must reconstruct the animal from a few bits of petrified bone dug out of rock. These essays on "The Tenth Generation" are an attempt in the direction of historical synthesis: to draw up a new and more satisfying picture of the process of history in ancient times, for which our only sources outside the Bible are the ancient records discovered and excavated during the past 100 years, and in particular the period in the history of Palestine, Lebanon, and Syria which immediately preceded the formation of that religious community which was called ancient Israel.

One of the great historians of the nineteenth century, Carlyle, once said, "A paradoxical philosopher carrying to the utmost length that aphorism of Montesquieu's 'Happy the people whose annals are tiresome,' has said

Note: A preliminary draft of this chapter appeared in *The Apprentice* [American University of Beirut] 2 (1971): 59–70.

1. The critical analysis of historiography and the purposes of historical writing is an enormous and perhaps impossible task. The essays by Plumb in *Death of the Past* illustrate this issue in most excellent fashion, and coincide to a considerable extent with my own conclusions.

'Happy the people whose annals are vacant.'"[2] It is very doubtful that the principle applies to ancient history, but it might very well apply in reverse to the period in which we are most interested—for the unhappiness, misery, and conflict mounting to a climax during the Late Bronze Age has left us a considerable documentation, which has never been adequately utilized historically.

For over a generation, archaeologists have divided archaeological ages in this region into Early Bronze 1–4, Middle Bronze 1–2, Late Bronze, and Iron 1–3. The boundary lines between these major periods are usually marked by ash layers—the permanent record of human violence[3] which are associated often with radical changes in the nature of archaeological objects found. For a hundred years or more, history has been explained as a constant process of barbarian invasions, of "fresh blood" invading and destroying older populations who had become corrupt and degenerate. Historians have thus carried out more genocides than has all humanity. This old theory is simply no longer satisfactory, far too simple, and most seriously misleading.

Part of the problem lies in the old myth of the "unchanging East." If according to the archaeological record something has changed, it must therefore follow, according to the old theory, that there is a new population, a new ethnic group. Though it is undeniable that from the dawn of history Palestine and Syria were the crossroads of civilization, trade, and migration, lands hospitable and almost always open to peoples from abroad, the idea that every change in pottery fashions must indicate a radical change in population must now be utterly rejected as historically naive. It would be just as absurd to argue that most of Western civilization became English because of the impressive argument from archaeological evidence for the mini-skirt, which everyone knows originated in London. Similarly, the argument of many archaeologists (in fact, this seems to be standard procedure) that differences in shapes of pottery proves differences in population groups simply ignores a constant feature of ancient and modern human life—that fashion is constantly changing.

It would seem that much of history has been produced by historians with a very deep-seated and therefore unconscious racial bias—and that bias is virtually without any basis in ancient history itself. The problem is a radically different one if we examine the ancient sources, what they say and what they do not say.

Those changes which seem to be attendant upon destruction levels in our archaeological sites are not necessarily caused by changes in the racial

2. *The French Revolution*, vol. I, bk. 11, chap. 1. Similar statements are found in Thomas Jefferson, George Eliot, and Schiller.

3. Schaeffer's attempt to account for the ash layer on the hypothesis that massive earthquakes were responsible is naive. Cf. above chap. I, n. 80.

make-up of the human occupants of the area. They occur with a sort of rhythmic frequency about every 250–300 years. The more radical destructions are often followed by what archaeologists may call an "Intermediate Period" or a "Dark Age," which merely means that for a century or two we have virtually no written documents from which we can reconstruct anything historical. Perhaps Carlyle was right in saying that such periods are the truly happy periods of human history, but I for one rather seriously doubt it, at least in our ancient periods.

Those destructions are separated from each other by a time span which corresponds very well to ten generations of humanity. I suggest that the Tenth Generation may well be worth more careful examination in the light of this rhythmic pattern of history in the past. How useful a concept it may be remains to be seen, but it may well serve to remind us that our own period is hardly unique—and that the historical process in the past has resulted in the destruction of seventy-five percent of human population within the period of a generation or two, and perhaps much less.[4]

The Tenth Generation witnessed and took part in the destruction of civilization. It seems possible—both from direct historical evidence and by analogy with later periods of crisis which seem to be similar in other respects —to make some generalizations about that Tenth Generation.

1. In the first place, it is not characterized by poverty. Real poverty and reduction to a subsistence-level economy are the features of the eleventh generation. Museum objects come from the period prior to destruction, not after. For the record, however, the museum objects come from palaces, temples, royal tombs, and the like—not from villages, whose small ruin mounds are hardly worth excavating from the point of view of the traditional archaeologist who wants museum objects: gold, semiprecious stones, works of art, and royal inscriptions.

2. Every indication leads to the conclusion that the Tenth Generation is characterized by a relatively high density of population, and it is worth considering that this may itself have something to do with the destruction. Zoologists have proven that even wild animals, when faced with over-population in a given area, often react with symptoms which in human beings we would call neurotic.[5]

4. Much more archaeological investigation into the question of fluctuations in population density is greatly needed, but the contrast between the high density of population in the Early Bronze Age and the following Intermediate Period seems beyond doubt at present.

5. J. B. Calhoun, "A 'Behavioral Sink'" in *Roots of Behavior*, ed. E. L. Bliss (New York, 1962), pp. 295–315. Phenomena include "pathological togetherness" as well as pathological aggression.

3. Perhaps most suggestive is a breakdown in confidence in the existing social and political organization. The inevitable result is increasing difficulty on the part of the organization in meeting its social functions. This is true not only of the political state, but of other kinds of social organization as well, including the family.[6]

4. There is some evidence for a rejection of past ideologies and a desire for something new. In fact, the Reformation in Europe is usually described in religious circles in this way—but the new is impossible until it happens. It is almost always a new look at the very old which brings about reform— but the important point is that the recent past and its ideology are rejected. Something else is felt to be necessary, desirable, and ultimately socially useful.[7]

5. With the rejection of the past and present, there is a loss of a sense of direction.[8] "When the past ceases to shed light on the future, the mind wanders in obscurity." There are no clear-cut criteria for deciding what is and is not important. There is little sense of immediate or long-range goals, only a desire to react to a situation which is felt to be intolerable. At the same time, there *may* be a heightened sense of ethical responsibility, especially since old formal obligations and conventional habits are rejected.[9]

6. There is an increasing resort to illegitimate force and violence, which is almost inevitable following the loss of confidence in the political organization which by definition is the "monopoly of force."[10]

7. If this is true, then the increased insecurity of everyone in the area is deeply felt, and there may be a sense of foreboding, of impending doom.[11]

6. It is most tempting to find evidence for the "generation gap" in the fourth commandment of the Decalogue ("honor thy father and thy mother"), which is otherwise so foreign to ancient Near Eastern patterns of social organization. Why should this relationship be placed under religious sanction? The context of a historical revolution can well supply the answer, as illustrated in the story of Gideon (Judg. 6:25-32). Since Gideon precedes the destruction of Shechem by at least one generation (Judg. 9), the event can hardly represent anything other than the community's transfer of allegiance from a local Baal to Yahweh—under the impact of a foreign aggression described in terms that come directly from the Amarna Letters (Judg. 6:3-4).

7. A new society based upon unification by covenant is precisely what Rib-Addi reports to be the political "conspiracy" of Abdu-Ashirta in *EA* 74.

8. The entire Amarna corpus of correspondence illustrates this thesis. As an illustration, one may cite the oft-repeated assertions of Rib-Addi that he is under pressure to join the movement of Abdu-Ashirta and his successor Aziru.

9. Cf. the noble aspiration to bring peace to all the lands in the propaganda message of Abdu-Ashirta (*EA* 74). It seems clear, however, that here, as usual, claims to high ethical goals are merely a means to political domination, or at least power status.

10. This is characteristic of ʿApiru organizations. For two very characteristic descriptions, compare *EA* 185-86 and 1 Sam. 27:6-12. In both cases, an ʿApiru group conducted raids upon villages outside the sovereignty of the immediate overlord.

11. See chap. VI and the repeated references to the destruction or loss of the lands there cited.

8. Doom comes: by whatever agency is perhaps by now irrelevant.[12] The political, social, and economic structures of society are destroyed or enormously weakened—usually by violence, either from within or without. Society simply disintegrates.[13]

9. Without such social and economic organization, deurbanization takes place. Cities are abandoned—sometimes permanently to the present day.[14] A high percentage of the population lose their lives, and what we call civilization—writing, art, political power, wealth, literature, large public buildings, and even institutionalized religion—either cease to exist, or disappear from view for a time to reemerge with the rebuilding process in much changed form. The Dark Age gradually comes into historical view once again. The new society may remember very little of the history of what preceded them, for that history has long been irrelevant. They conclude that God destroyed the old in a Great Flood.[15]

This is consciously and admittedly a hypothetical model of the Tenth Generation, but the goal of the process, which we see in the ash layers and Dark Ages of the Bronze Age, is not hypothetical, nor is it an inevitable one. It is such periods that have also produced some of the most permanent changes in human society. The rise of Christianity anticipated the Tenth Generation by one generation—and it is exactly 240 years from the outbreak of the Maccabean violence until the final destruction of Jerusalem. Many more examples of this Tenth Generation crisis can easily be given, but it is not the present purpose to prove that this historical analysis is correct and accurate history. Rather, this dramatizes the fact that all humanity since the beginning of history has been faced repeatedly with the very serious problem of how to resolve a most basic conflict in human society. That conflict can be very simply described. Large social organizations *seem* at least to be absolutely necessary to support a large population,[16] but large social organizations rapidly tend to become intolerable to the populations, the real persons whom they dominate and whose well-being they effectively determine, whether for good or evil.

12. That the destruction is caused in part by *internal* agencies is demonstrated by Judg. 9 and reinforced by the letter from the military official of Cyprus to the king of Ugarit in *Ugaritica*, 5 (RS 20.18): 83–84. The universal destruction that attended the end of the Late Bronze Age cannot be attributed to one simple cause.

13. Perhaps the destruction levels may be best interpreted as evidence that in revolutionary movements it is always the barbarians who take over—for they are the most adequately equipped for the exercise of violence, and the least inhibited.

14. Ugarit itself is the parade example.

15. Greek traditions particularly seem to lend themselves to this interpretation of the folkloristic meaning of Flood stories.

16. V. Gordon Childe's observations of the relationship between human technology and population density may well illustrate this thesis (*What Happened in History* [1942], pp. 37, 59, 85, 165, 269, 270). It should be clear, however, that "state socialism" is no more Oriental than it is Western or primitive.

Wherever we look in the ancient world we find a very complex group of larger social organizations than the basic unit, the village and its families. But we have still not even mentioned the largest unit which we call empire. All of this should, if really taken seriously, demonstrate the complete fallacy of speaking of ancient peoples as races or ethnic groups, as modern scholars do with such facility.[17] So Germans speak of *das assyrische Volk* and other biblical scholars speak of the Moabites, Edomites, Canaanites, and the like as though we were dealing with some mystical racial group. This simply is not true, for terms of this sort do not designate a homogeneous population group which contrasts to another in the area. The terms designate various kinds of larger social organizations, especially political ones—and, as we have seen, all such large units are ephemeral in the ancient world. One might add, in the modern world as well, for no modern state has existed in its present form for more than two and a half centuries. It should go without saying that even England as a state is something radically different from what it was in the days of King George III, even though there is a continuity of cultural tradition, population, and political organization. But even it was composed originally of Angles, Saxons, Normans, Celts, Welsh, and who knows what else? There is no such things as a "pure" ethnic group in any historical social organization of considerable extent, and such racist ideas must be expunged from the conceptual baggage of the historian. It is most probable that no such concept even existed in the ancient world.

The Tenth Generation, then, is one which witnessed, and perhaps caused, the destruction of the large social organizations of its time. All over the ancient world, empires and states collapsed, and our historical sources and written documents cease entirely or are radically reduced in number for two or three centuries. With very few exceptions, every excavated site from Palestine and Syria to Greece shows the remains of destruction by violence during the period from 1250 to 1150 B.C. The old empires of Mitanni, Arzawa, Kizzuwatna, the Hittites, the Myceneans, the Minoans, and many others ceased to exist. Many were completely forgotten until exhumed by the archaeologist and historian, and others lived on only as garbled legends, folk tales, and myths. There were migrations and population movements, particularly, it would seem, from north to south and from east to west. Transjordan, an area which seems to have had previously a very meager population, saw a rapid rise in population density at about this time.

17. A feeling of ethnic unity is a product of long-continued proximity and functional relationships, as Service has pointed out (*PSO*, p. 176), but the role of political organization as the important factor could hardly be described in a work on primitive cultures.

Formerly, scholars summarized the historical event as follows: "The Canaanites suffered a terrible blow in the early Iron Age. In the North, the Arameans drove them out of the area from Damascus to Aleppo, on the coast, the Phoenicians drove them out, and in Palestine, the ancient Israelites. The Canaanites lost nine-tenths of their territory." Today, this can be seen as complete nonsense. Let us look at the situation before and after the destruction, focusing our attention on Palestine and the Mediterranean Coast, not only because we have the most evidence from these regions, but also because what happened in Palestine three thousand years ago has enormous impact on history today.

In the mid-fourteenth century B.C., all of Palestine and Syria was divided into a multitude of city-states, with the surrounding villages rather typical of the Roman *pagi* which we have described.[18] Most of those city-states were ruled by a king, and a very high percentage of those kings bore names of foreign origin. They were Hurrians, probably from east central Anatolia, Indo-Iranians, Hittites, even Indo-Aryans. In other words, they were in all probability very recent newcomers who had moved in with superior weapons and military organization and established small military states. Virtually all of these were constantly attempting to enlarge their territories at the expense of the nearest neighbor, and in the process they were seriously undermining the Egyptian peace which had long exercised at least nominal control as far as the Euphrates River. The Egyptian Empire was also increasingly challenged by the Hittites to the north. Some of the older dynasts seem to have maintained themselves on the throne, but they seem, like pitiful Rib-Addi of Byblos, to have been absolutely incapable of defending themselves against the power and unscrupulous tactics of their ambitious neighbors. Rib-Addi pleads unceasingly for a detachment of fifty Egyptian soldiers to keep the peace and maintain the empire. He says, "Do you not know the land of Amurru, that they follow the one who has power?"[19]

And well they might, for the favorite tactic of the unscrupulous and ambitious kinglets was to harass, threaten, and drive out the villagers, who therefore could not till their fields and tend their grapevines, fig trees, and olive orchards. Not only were the villagers deprived of their livelihood and food supply, the nearby city found its food supply cut off as well. Rib-Addi again had to plead desperately to the pharaoh to send food for the inhabitants of his city, who were living on meagerly rationed supplies, verging on starvation, and about to revolt. The enemy was circulating messages to the local populations saying, in effect: "Assassinate your rulers, join us in a

18. Chap. VII.
19. *EA* 73:14–16.

covenant, then you will have peace." For most of the population, particularly the productive villagers, it seemed to make little difference which side won. We cannot be absolutely sure, but we have proof for some areas that wherever a new king, a new military conqueror, assumed power, he regarded the whole of the productive land as his personal property to dispose of as he saw fit—to the military aristocracy upon whom his power depended, to his bureaucrats, and probably to the temple priests who were also part of the aristocracy. The villagers who did the work could not own their own land —they were merely permitted to till it and were then generously given enough of their own produce to keep them alive. They were the *ḫupšu*, who were probably not legally slaves but might as well have been.[20]

Under such circumstances, it is not surprising that the king and his bureaucracy had a surplus of funds sufficient to support an ambitious building program—of temples, palaces, fortifications, a sumptuous and expensive art in ivory, gold, and silver. His royal glory had to be magnificent to convince the peasants that they should be proud to have such a king. His prestige was based on divine right; his power derived from the gods whose rituals he supported both in the building of elaborate temples and in the maintenance of a complex priestly organization, and of course from an army and often a navy.[21] It seems probable that as in early Rome, almost any eminent citizen would likely end up being ordained as a priest in addition to his regular position as a professional military or as some important official in the royal bureaucracy. We have one text which proves this, but we cannot be sure that this was the normal pattern.[22]

We can see, then, the far-reaching consequences of the harassment of the villagers. The economic system upon which the whole kingdom depended, its prestige, authority, military power, and even religion evaporated—but its place was taken usually by another example of the same system. The most fundamental problem, I suggest, was ideological: no political power seems to have had any concept of an ideology other than that of the divine delegation of power to the king. The king owned all productive resources of the kingdom; he exercised the monopoly of force through a complex bureaucracy and an effective professional military group, both of whom made up the aristocracy. This, it seems to me, might well be termed "primitive com-

20. I. Mendelssohn, "The Canaanite Term for 'Free Proletarian,'" *BASOR* 83 (1941): 36–39. Cf. also *FSAC*, p. 285.

21. Plumb could not be more wrong in his assertion that religion was used to reconcile the peasants to agrarian life. It was used to reconcile them to the idea that most of their produce should be seized without compensation to support the royal establishment, its bureaucracies, armies, and priesthoods (*Death of the Past*, pp. 26–27). The rationalizations of the process differ widely in different times.

22. *AT* 15 (p. 39).

munism," just as the tribal system of a much earlier period has been termed, I believe justly, "primitive democracy."[23]

If the fundamental problem was ideological—the political theology which we tend to discuss as "religion" these days—it is no wonder that the process has the same result all over the Eastern Mediterranean: destruction. The fundamental solution was also ideological, and it is this which originally constituted the biblical faith, the introduction of monotheism, and the elevation of ethic to the place of preeminent concern. If the Old Testament faith has been condemned, vilified, radically distorted, and rejected, all this merely furnishes grounds for understanding the meaning of the New Testament prayer, "hallowed be Thy Name."

First, let us take a better look at Bronze Age religious ideology. The gods represent those factors which are not really under the control of human beings, but upon which human well-being depends. Fundamental, of course, is fertility. The fertility cult is not merely a license for sexual orgies, though such rituals must have existed under certain circumstances at least.[24] The fertility cult is merely the recognition of the extreme dependence of man upon the whole process of tilling, planting, and harvesting crops, the techniques of which were well known and understood, but the success of which was dependent upon factors beyond human control—rainfall, absence of disease and locusts, and the miraculous process itself of growth and maturation of the planted seed. The fertility cult is the deification of the process of production and, appropriately, is usually represented as a Great Mother and a god of the storm who brings the fertilizing rainfall.

The god of the storm, however, is not merely that. He is the god also of power—the executive deity who delegates power to the king. I maintain that the original and extremely old fertility cult has in all our written sources been politicized—that is, the old beliefs and myths have been readapted to show that the whole mythical process of fertility and power, the complex of powers and forces beyond the control of man, have actually been delegated to the king who exercises them as the chosen of the gods. Therefore, any opposition to the king not only constitutes a political crime: It constitutes also an offense against the gods and endangers the whole system of production and maintenance of peace. The old cult was therefore not only mythically true, it was historically true—and remains so to the present day. The question was not, however, whether the maintenance of the system of production and of political authority was dependent on factors beyond the control of man—the question was whether that system could operate in a

23. C. U. Wolf, "Traces of Primitive Democracy in Ancient Israel," *JNES* 6 (1947): 98–108.

24. Chap. IV and literature there cited.

way which was tolerable to the human beings living under it. It is here that it failed—and its gods as well.

Second, with this background, it is rather easy to see the points of opposition in the early biblical faith. The king, who was the focus of the whole political, economic, and religious system, was eliminated. If he ruled by delegated authority from the gods, why did not the God himself rule? That is exactly what ancient Israel was—the Kingdom of God. There was no delegation of power to a centralized political system. The deification of productivity and power were rejected in favor of an ethic, and economic prosperity and security were presented as byproducts of ethical obedience rather than as evidences of divine authority and proof of divine favor. The totally evil could prosper too; in fact, it is considerably easier to be wealthy if ethical obligations to fellow man do not interfere. After all, ethic is based on an ability to predict the consequences of behavior and to avoid patterns of behavior which create conflict. The rejection of ethic, then, is the rejection of peace. This is precisely what happened with the reestablishment of monarchy under Kings David and Solomon, and the consequences were similar to those of the Late Bronze Age monarchies. Though much limited, and somewhat chastened, the kingship continued for another century or two, and toward the end the Old Testament prophets reaffirmed the predictive validity of ethic—a state which sacrifices ethical obedience for temporary gain must be destroyed by God Himself. This is the gist of biblical prophecy; it was reaffirmed by Jesus Himself.

Who, then, were the biblical Israelites? It should first be pointed out that the term *Yiśreʾēlī* does not even occur in the early parts of the Bible, except in one passage which all scholars have agreed is a textual error. The gentilic occurs in one passage which is generally regarded as very late.[25] It is a confusion in terminology to speak of the "Israelites" as an ethnic group during the biblical period. Israel is the designation of a religious community, of a large social organization, that constituted the Kingdom of God. It consisted of twelve, more or less, tribes, which were almost certainly administrative districts each under a *nāśīʾ*. The very word "tribe" in biblical Hebrew means "staff," which is, in the ancient world, a symbol of authority. The twelve tribes were comprised of those members of the population of Palestine and Transjordan who had accepted the rule of God. This constitutes the only perceptible difference between them and the non-Yahwistic population, which tended to center in the old Canaanite city-states that Israel did not convert to Yahwism and which it had neither the motivation nor the military power to conquer until the reign of David.

It follows, then, that accepted traditions about the so-called Conquest

25. See chap. VII, on Lev. 24:10–11.

have missed the point. True it was that the "seven nations" were driven out —but those seven nations designate not ethnic groups and mass populations, but political regimes of foreign military conquerors together with their fighting forces. Of the seven nations in the biblical list, the Hittites, Hivvites, Girgashites, and Perizzites can be identified, either probably or with certainty, as recent invaders from Anatolia. The Jebusites have a name most closely associated with northern and northeastern Syria.[26] Of the biblical clan and personal names, there is a fairly impressive sprinkling of Anatolian names. The biblical traditions themselves insist upon the fact that a "mixed multitude" went out of Egypt with Moses, which means that they were *not* characterized by an ethnic unity. The tribe of Judah is said to be part Canaanite in origin, and the Transjordanian half-tribe of Machir is said to be Aramean. Ethnic identity or unity and descent from a common ancestor had nothing to do with the nature of early Israel. The Kenizzite who is supposed to have been disinherited from the land in the narrative of the promise to Abraham is certainly the same Kenaz whose "son" Othniel took Kiryat-Sefer in chapter 1 of Judges. Biblical names include Hittite, Hurrian, Egyptian, Libyan, Amorite, Canaanite, proto-Aramean, probably Indo-Iranian, and other names, just as do our name lists from Late Bronze Age cities. The conflict between ancient Israel and the non-Israelite population had nothing to do with ethnic identity. It was a conflict with an old political regime or system of regimes which were rightly dying out all over the civilized world because they valued power more than ethic, and valued property and wealth more than persons.

There was chaos, conflict, war, but of one thing we can be absolutely certain. Ancient Israel did not win because of superior military weapons or superior military organization.[27] It did not drive out or murder en masse whole populations. The gift of the land meant merely that the old political regimes and their claim to ownership of all land was transferred to God Himself, and it was not given to Israel, for ancient Israel did not own it.[28] It is for this reason that sale of land is impossible in biblical law—one cannot sell something one does not own. Enjoyment of the land, then, is absolutely dependent upon obedience to the sovereign who does own it—and it is for

26. The change of name from Jerusalem to Jebus (Judg. 19:10-11) must have taken place between ca. 1370 (Amarna still uses the old name) and 1200—and would then have been a result of the inundation of sea-peoples that took place during this period, as argued above in chap. V.

27. This has been argued convincingly by Glock in "Warfare in Mari and Ancient Israel."

28. The story of Naboth's vineyard is a case in point. The post-Exilic prophecy of Ps. 50:10-12 illustrates the normative concept.

this reason that the destruction and expulsion from the land took place, not once but twice.

It was not until almost a thousand years after Moses that the religious community finally settled on the idea that ethnicity or race was the foundation of the religious community and the basis of individual identity, when Ezra and Nehemiah forced the divorce of non-Jewish wives.[29] Four hundred years later a new reformation took place which proclaimed that the Kingdom of God could not be identified with any political state, with any systematic use of law and force, and, above all, it could not be identified with any ethnic group.[30] This, the new Israel of God, the New Testament church, is the fulfilment of biblical prophecy. To admit anything else as that fulfilment is to turn the clock back to the Late Bronze Age and worse, for at least that period knew nothing of modern doctrines of racial superiority.

The Tenth Generation is always faced with the alternatives of deification of force or the deification of ethic. The probable consequences of the two ways should be reasonably predictable.[31]

29. Ezra 9–10; Neh. 13:23–30.
30. Matt. 8:11–12; Luke 3:8 (John the Baptist); Gal. 3:28.
31. For a belated illustration of a similar thesis in an entirely unrelated field, see now the demonstration of Jay Wright Forrester that the life cycle of a modern industrial city is about 250 years and that many of the political measures taken to remedy social ills simply make them worse (*Urban Dynamics* [Cambridge, Mass., 1969]).

BIBLIOGRAPHY

Albright, William F. *Archaeology and the Religion of Israel.* 3d ed. Baltimore, 1953.
———. *From the Stone Age to Christianity.* 2d ed. rev. Baltimore, 1957.
———. *Vocalization of the Egyptian Syllabic Orthography.* New Haven, 1934.
Archives royales de Mari (traduit). Paris, 1942—.
Barr, James. *Comparative Philology and the Text of the Old Testament.* Oxford, 1968.
Bauer, Theo. *Die Ostkanaanäer: Eine philologisch-historische Untersuchung über die Wanderschicht der sogenannten "Amoriter" in Babylonien.* Leipzig, 1926.
Bess, S. Herbert. "Systems of Land Tenure in Ancient Israel." Ph.D. dissertation, University of Michigan, 1963.
The Biblical Archaeologist Reader. 3 vols. Garden City, N.Y., 1961–70.
Blackman, A. M., and Fairman, H. W. "The Myth of the Horus at Edfu," *JEA* 21 (1935): 26–36; 28 (1942): 32–38; 29 (1943): 2–36; 30 (1944): 5–22.
Bossert, H. T. *Altsyrien.* Tübingen, 1951.
———. *Ein hethitisches Königssiegel.* Istanbuler Forschungen, no. 18. Berlin, 1944.
Bright, John. *Jeremiah.* Anchor Bible. Garden City, N.Y., 1965.
Callaghan, R. T. *Aram Naharaim.* Rome, 1948.
Cardozo, B. N. *Selected Writings.* Edited by M. Hall. New York, 1947.
Cazelles, H. *Études sur le code de l'alliance.* Paris, 1946.
Delitzsch, Friedrich. *Sumerisches Glossar.* Leipzig, 1914.
De Vaux, Roland. *Ancient Israel: Its Life and Institutions.* New York, 1961.
Dull, R. "Vom Vindex zum Index," *Zeitschrift der Savigny Stiftung für Rechtsgeschichte,* 54 (1934): 98–136; 55 (1935): 9–35.
Dunlop, D. M. *A History of the Jewish Khazars.* Princeton, N.J., 1954.
Dussand, René. *Les origines cananéennes du sacrifice israélite.* Paris, 1st ed., 1921; 2d ed., 1941.
Edelstein, E. J., and Edelstein, L. *Asclepius.* Baltimore, 1945.
Erman, A., and Grapow, H. *Ägyptisches Handwörterbuch.* Berlin, 1921.
Falkenstein, A. *Haupttypen der sumerischen Beschwörung.* Leipzig, 1931.
Fontenrose, J. *Python.* Berkeley, 1959.
Frankfort, Henri. *Cylinder Seals.* London, 1939.
———, et al. *The Intellectual Adventure of Ancient Man.* Chicago, 1946.
Friedrich, J. *Hethitisches Elementarbuch.* Vol. II. Heidelberg, 1946.
Frye, Richard N. *The Heritage of Persia.* Cleveland, 1963.
Gardiner, Alan. *Ancient Egyptian Onomastica.* 3 vols. Oxford, 1946.
———. "Horus the Behdetite," *JEA* 30 (1944): 23–60.

Gelb, Ignace J.; Purves, Pierre M.; and MacRae, Allan A. *Nuzi Personal Names.* Oriental Institute Publications, no. 57. Chicago, 1943.

Gershevitch, I. *The Avestan Hymn to Mithra.* Cambridge, 1959.

Glock, A. "Warfare in Mari and Early Israel." Ph.D dissertation, University of Michigan, 1968.

Glueck, Nelson. "Explorations in Eastern Palestine." In *Annual of the American Schools of Oriental Research,* vols. 14, 18, 19, 25–28. New Haven, 1934–51.

Goetze, A. "Hattusilis," *MVAG* 29 (1925).

———. *Kleinasien.* 2d ed. Munich, 1957.

———. "The Linguistic Continuity of Anatolia as Shown by Its Proper Names," *JCS* 8 (1954).

Gordon, Cyrus H. *Ugaritic Textbook.* Rome, 1965.

Greenberg, Moshe. *The Ḫab/piru.* American Oriental Series, vol. 39. New Haven, 1955.

Gröndahl, Frauke. *Die Personennamen der Texte aus Ugarit.* Rome, 1967.

Harper, R. F. *Assyrian and Babylonian Letters.* Chicago, 1892–1914.

Hencken, H. *Tarquinia, Villanovans and Early Etruscans.* Cambridge, Mass., 1968.

Hillers, Delbert R. *Covenant: The History of a Biblical Idea.* Baltimore, 1969.

———. *Treaty Curses and the Old Testament Prophets.* Biblica et Orientalia, no. 16. Rome, 1964.

Holmes, Oliver Wendell. *The Common Law.* Boston, 1891.

Hopkins, Clark. "The Canopy of Heaven and the Aegis of Zeus," *Bucknell Review,* December, 1964, pp. 1–16.

Houwink Ten Cate, Ph. H. J. *The Luwian Population Groups of Lycia and Cilicia Aspera during the Hellenic Period.* Leiden, 1961.

Huffmon, Herbert B. *Amorite Personal Names in the Mari Texts.* Baltimore, 1965.

Ingholt, Harald. *Rapport préliminaire sur . . . fouilles à Hama.* Copenhagen, 1940.

The Interpreter's Dictionary of the Bible. Edited by G. A. Buttrick. 4 vols. Nashville, 1962.

Jung, C., and Kerényi, C. *Essays on a Science of Mythology.* New York, 1949.

Kantorowicz, Ernst H. *The King's Two Bodies.* Princeton, 1967.

Kaufman, Y. *The Religion of Israel.* Translated and abridged by M. Greenberg. Chicago, 1960.

Knudtzon, J. A.; Weber, Otto; and Ebeling, Erich. *Die El-Amarna-Tafeln.* Vorderasiatische Bibliothek, no. 2. Leipzig, 1911.

Koschaker, P. *Babylonisch-assyrisches Bürgschaftsrecht.* Leipzig, 1911.

Kraus, H.-J. *Geschichte der historisch-kritischen Erforschung des alten Testaments.* Neukirchen, 1956.

Lane, E. W. *An Arabic-English Lexicon.* 8 vols. London, 1863–85.

Laroche, E. *Dictionnaire de la langue louvite.* Paris, 1959.

———. *Les hiéroglyphes hittites.* Paris, 1960.

———. *Les noms des Hittites.* Paris, 1966.

———. "Récherches sur les noms des dieux hittites," *Revue hittite et asianique,* 46 (1946–47).

Laumonier, A. *Les cultes indigènes en Carie.* Paris, 1958.

Lidzbarski, Mark. *Ephemeris für semitische Epigraphik.* 3 vols. Giessen, 1902–15.

Littman, Enno. *Safaitic Inscriptions.* Publication of the Princeton University Expeditions to Syria, division IV, section C. Leiden, 1943.

Luckenbill, D. D. *Ancient Records of Assyria and Babylonia.* 2 vols. Chicago, 1926–27.

Luke, J. Tracy. "Pastoralism and Politics in the Mari Period: A Re-examination of the Character and Political Significance of the Major West Semitic Tribal Groups on the Middle Euphrates, ca. 1828–1758." Ph.D. dissertation, University of Michigan, 1965.

Mallowan, Max. *Nimrud and Its Remains.* 3 vols. London, 1966.

Meisel, J. *Counterrevolution: How Revolutions Die.* New York, 1966.

Mercer, S. A. B. *The Tell El-Amarna Tablets.* Toronto, 1966.

Merz, Erwin. *Die Blutrache bei den Israeliten.* Beiträge zur Wissenschaft vom Alten Testament, vol. 20. Leipzig, 1916.

Middleton, J. *Lugbara Religion.* London, 1960.

Milgrom, J. "The Shared Custody of the Tabernacle and a Hittite Analogy," *JAOS* 90 (1970).

Neumann, G. *Untersuchungen zum Weiterleben hethitischen und luwischen Sprachgutes in hellenistischer und römischer Zeit.* Wiesbaden, 1961.

Noth, Martin. *The History of Israel.* 2d ed. New York, 1963.

Oppenheim, A. Leo, et al., eds. *The Assyrian Dictionary of the Oriental Institute of the University of Chicago.* Chicago, 1956—.

Otten, H. "Ritual bei Erneuerung von Kultsymbolen hethitischer Schutzgottheiten." In *Festschrift Johannes Friedrich.* Heidelberg, 1959.

Oxtoby, W. G. *Some Inscriptions of the Safaitic Bedouin.* New Haven, 1968.

Pallis, S. A. *The Babylonian Akîtu Festival.* Copenhagen, 1926.

Palmer, L. R. *The Interpretations of Mycenean Greek Texts.* Oxford, 1963.

Paulys Real-Encyclopädie der classischen Altertumswissenschaft. Edited by G. Wissowa. Stuttgart, 1894—.

Plumb, J. H. *The Death of the Past.* Boston, 1970.

Posener, Georges. *Princes et Pays d'Asie et de Nubie: Textes hiératiques sur des figurines d'envoûtement du Moyen Empire.* Brussels, 1940.

Pritchard, J. B., ed. *Ancient Near Eastern Texts relating to the Old Testament.* 2d ed. Princeton, 1955.

———. *The Ancient Near East in Pictures.* Princeton, 1954.

Schaeffer, Claude F.-A. *Stratigraphie comparée et chronologie de l'Asie occidentale.* London, 1948.

———. *Ugaritica.* Paris, 1968.

Service, E. R. *Primitive Social Organization.* New York, 1962.

Smith, W. S. *A History of Egyptian Sculpture and Painting in the Old Kingdom.* London, 1946.

Tussman, J. *Obligation and the Body Politic.* New York, 1960.

Von Soden, Wolfram, ed. *Akkadisches Handwörterbuch: Unter Benutzung des lexikalischen Nachlasses von Bruno Meissner (1868–1947).* Wiesbaden, 1959—.

Weippert, M. *Die Landnahme der israelitischen Stämme.* Göttingen, 1967. English translation: *The Settlement of the Israelite Tribes in Palestine: A Critical Survey of Recent Scholarly Debate.* Studies in Biblical Theology, series 2, no. 21. Naperville, Ill., 1972.

Wilson, J. *The Burden of Egypt.* Chicago, 1951.

Wiseman, D. J. *The Alalakh Tablets.* London, 1953.

Wright, G. E. *Biblical Archaeology.* Philadelphia, 1957.

————. *Shechem: The Biography of a Biblical City.* New York, 1965.

————, ed. *The Bible and the Ancient Near East: Essays in Honor of W. F. Albright.* Garden City, 1961.

Zgusta, L. *Kleinasianische Personennamen.* Prague, 1964.

INDEX OF
BIBLICAL CITATIONS

INDEX OF

SUBJECTS AND AUTHORS

THE JOHNS HOPKINS UNIVERSITY PRESS

This book was composed in Bembo text and Centaur
display type by William Clowes & Sons, Ltd.
from a design by Victoria Dudley.
It was printed by Universal Lithographers, Inc. on
S. D. Warren's 60-lb. Sebago paper, in a text shade.
The book was bound by L. H. Jenkins, Inc.
in Columbia Mills' Llamique cloth.